BIO-PESTICIDE AND INTEGRATED PEST MANAGEMENT

BIO-PESTICIDE AND INTEGRATED PEST MANAGEMENT

Dr. D.V. Bhagat

RANDOM

RANDOM PUBLICATIONS

NEW DELHI - 110 002 (INDIA)

Bio-Pesticide and Integrated Pest Management

ISBN 978-93-51116-77-6

Published in 2015 in India by

RANDOM PUBLICATIONS

4376-A/4B, Gali Murari Lal, Ansari Road
New Delhi-110 002
Phone: +9111-43580356, 23289044
E-mail: randomexports@gmail.com; sales@randompublications.com; info@randompublications.com

Reprinted 2024

Type Setting by: Friends Media, Delhi-110089
Digitally Printed at: Replika Press Pvt. Ltd.

Preface

Biopesticides are certain types of pesticides derived from such natural materials as animals, plants, bacteria, and certain minerals. For example, canola oil and baking soda have pesticidal applications and are considered biopesticides. In 2014, there are more than 430 registered biopesticide active ingredients and 1320 active product registrations. Biopesticides fall into three major classes. Microbial pesticides consist of a microorganism (e.g., a bacterium, fungus, virus or protozoan) as the active ingredient. Microbial pesticides can control many different kinds of pests, although each separate active ingredient is relatively specific for its target pests. For example, there are fungi that control certain weeds, and other fungi that kill specific insects. The most widely used microbial pesticides are subspecies and strains of Bacillus thuringiensis, or Bt. Each strain of this bacterium produces a different mix of proteins, and specifically kills one or a few related species of insect larvae. While some Bt's control moth larvae found on plants, other Bt's are specific for larvae of flies and mosquitoes. The target insect species are determined by whether the particular Bt produces a protein that can bind to a larval gut receptor, thereby causing the insect larvae to starve. Plant-Incorporated-Protectants (PIPs) are pesticidal substances that plants produce from genetic material that has been added to the plant. For example, scientists can take the gene for the Bt pesticidal protein, and introduce the gene into the plant's own genetic material. Then the plant, instead of the Bt bacterium, manufactures the substance that destroys the pest.

Biochemical pesticides are naturally occurring substances that control pests by non-toxic mechanisms. Conventional pesticides, by contrast, are generally synthetic materials that directly kill or inactivate the pest. Biochemical pesticides include substances, such as insect sex pheromones, that interfere with mating, as well as various scented plant extracts that attract insect pests to traps. Biopesticides for use against crop diseases have already established themselves on a variety of crops. For example, biopesticides already play an important role in controlling downy mildew diseases. Their benefits include: a 0-Day Pre-Harvest Interval, the ability to use under moderate to severe disease pressure, and the ability to use as a tank mix or in a rotational programme with other registered fungicides. Because some market studies estimate that as much as 20% of global fungicide sales are directed at downy mildew diseases, the integration of biofungicides into grape production has substantial benefits in terms of extending the useful life of other fungicides, especially those in the reduced-risk category. A major growth area for biopesticides is in the area of seed treatments and soil amendments. Fungicidal and biofungicidal seed treatments are used to control soil borne fungal pathogens that cause seed rots, damping-off, root rot and seedling blights. They can also be used to control internal seed–borne fungal pathogens as well as fungal pathogens that are on the surface of the seed. Many biofungicidal products also show capacities to stimulate plant host defence and other physiological processes that can make treated crops more resistant to a variety of biotic and abiotic stresses. Integrated pest management (IPM), also known as Integrated Pest Control (IPC) is a broad-based approach that integrates practices for economic control of pests. IPM aims to suppress pest populations below the economic injury level (EIL).

The book goes on to deal with the control methods: biological, cultural and chemical. The biological control agents discussed include both native and introduced species that attack pests. This book, intended for all those involved in studying entomology, crop protection and pest management.

I thank all members of my team who have helped in the preparation of the book. My special thanks go to "Random Publications" who have published the book.

— *Dr. D.V. Bhagat*

Contents

Chapter 1

Introduction

Biopesticides, a contraction of 'biological pesticides', include several types of pest management intervention: through predatory, parasitic, or chemical relationships. The term has been associated historically with biological control - and by implication - the manipulation of living organisms. Regulatory positions can be influenced by public perceptions, thus:

- in the EU, biopesticides have been defined as "a form of pesticide based on micro-organisms or natural products".
- the US EPA states that they "include naturally occurring substances that control pests (biochemical pesticides), microorganisms that control pests (microbial pesticides), and pesticidal substances produced by plants containing added genetic material (plant-incorporated protectants) or PIPs".

They are typically created by growing and concentrating naturally occurring organisms and/or their metabolites including bacteria and other microbes, fungi, nematodes, proteins, *etc.* They are often considered to be important components of integrated pest management (IPM) programmes, and have received much practical attention as substitutes to synthetic chemical plant protection products (PPPs).

Biopesticides fall into three major classes:

- Microbial pesticides which consist of bacteria, entomopathogenic fungi or viruses (and sometimes includes the metabolites that bacteria or fungi produce). Entomopathogenic nematodes are also often classed as microbial pesticides, even though they are multi-cellular.

- Biochemical pesticides or herbal pesticides are naturally occurring substances that control (or monitor in the case of pheromones) pests and microbial diseases.
- Plant-incorporated protectants (PIPs) have genetic material from other species incorporated into their genetic material (*i.e.* GM crops). Their use is controversial, especially in many European countries.

Biopesticides have usually no known function in photosynthesis, growth or other basic aspects of plant physiology; however, their biological activity against insect pests, nematodes, fungi and other organisms is well documented. Every plant species has developed a built-in unique chemical complex structure that protects it from pests. The plant kingdom offers a diverse array of complex chemical structures and almost every imaginable biological activity. These biodegradable, economical and renewable alternatives are used especially under organic farming systems.

Examples: *Bacillus thuringiensis*, a bacterial disease of Lepidoptera, Coleoptera and Diptera, is a well-known insecticide example. The toxin from *B. thuringiensis* (Bt toxin) has been incorporated directly into plants through the use of genetic engineering. The use of Bt Toxin is particularly controversial. Its manufacturers claim it has little effect on other organisms, and is more environmentally friendly than synthetic pesticides. However, at least one scientific study has suggested that it may lead to slight histopathological changes on the liver and kidneys of mammals with Bt toxin in their diet.

Other microbial control agents include products based on:

- entomopathogenic fungi (*e.g. Beauveria bassiana, Lecanicillium* spp., *Metarhizium* spp.),
- plant disease control agents: include *Trichoderma* spp. and *Ampelomyces quisqualis* (a hyper-parasite of grape powdery mildew); *Bacillus subtilis* is also used to control plant pathogens.
- beneficial nematodes attacking insect (*e.g. Steinernema feltiae*) or slug (*e.g. Phasmarhabditis hermaphrodita*) pests
- entomopathogenic viruses (*e.g.. Cydia pomonella* granulovirus).
- weeds and rodents have also been controlled with microbial agents.

Various naturally occurring materials, including fungal and plant extracts, have been described as biopesticides. Products in this category include:

- Insect pheromones and other semiochemicals
- Fermentation products such as Spinosad (a macro-cyclic lactone)
- Chitosan: a plant in the presence of this product will naturally induce systemic resistance (ISR) to allow the plant to defend itself against disease, pathogens and pests.
- Biopesticides may include natural plant-derived products, which include alkaloids, terpenoids, phenolics and other secondary chemicals. Certain vegetable oils such as canola oil are known to have pesticidal properties. Products based on extracts of plants such as garlic have now been registered in the EU and elsewhere.
- Naturally occurring minerals such as baking soda may also have pesticidal applications.

Applications

Biopesticides are typically microbial biological pest control agents that are applied in a manner similar to chemical pesticides. In order to implement these environmentally friendly pest control agents effectively, it can be important to pay attention to the way they are formulated and applied.

Biopesticides for use against crop diseases have already established themselves on a variety of crops. For example, biopesticides already play an important role in controlling downy mildew diseases. Their benefits include: a 0-Day Pre-Harvest Interval, the ability to use under moderate to severe disease pressure, and the ability to use as a tank mix or in a rotational programme with other registered fungicides. Because some market studies estimate that as much as 20% of global fungicide sales are directed at downy mildew diseases, the integration of biofungicides into grape production has substantial benefits in terms of extending the useful life of other fungicides, especially those in the reduced-risk category.

A major growth area for biopesticides is in the area of seed treatments and soil amendments. Fungicidal and biofungicidal seed treatments are used to control soil borne fungal pathogens that cause seed rots, damping-off, root rot and seedling blights. They can also be used to control internal seed–borne fungal pathogens as well as fungal pathogens that are on the surface of the seed. Many biofungicidal products also show capacities to stimulate plant host defences and other physiological processes that can make treated crops more resistant to a variety of biotic and abiotic stresses.

Advantages

- Harmful residues not detected
- Can be cheaper than chemical pesticides when locally produced.
- Can be more effective than chemical pesticides in the long-term (as demonstrated, for example, by the LUBILOSA Programme)
- Biodegradable

Disadvantages

- High specificity: which may require an exact identification of the pest/pathogen and the use of multiple products to be used; although this can also be an advantage in that the biopesticide is less likely to harm species other than the target
- Often slow speed of action (thus making them unsuitable if a pest outbreak is an immediate threat to a crop)
- Often variable efficacy due to the influences of various biotic and abiotic factors (since biopesticides are usually living organisms, which bring about pest/pathogen control by multiplying within the target insect pest/pathogen)
- Living organisms evolve and increase their resistance to biological, chemical, physical or any other form of control. If the target population is not exterminated or rendered incapable of reproduction, the surviving population can acquire a tolerance of whatever pressures are brought to bear, resulting in an evolutionary arms race.

Bioherbicide

A bioherbicide is a biologically based control agent for weeds. Among the three major types of pesticides (agricultural pest-control agents) herbicides are used to control weeds, or undesirable plants. (The other major classes of pesticides are insecticides, which control insects; and fungicides, which control fungi. Normally, any pesticide can only be expected to control, but not eradicate, agricultural pests.)

Almost every agricultural pest has at least one naturally occurring enemy that will reduce its population. Bioherbicides are weed-control agents that utilize such naturally occurring enemies, rather than depending on man-made chemicals. This can be important because agents of biological control ordinarily have many fewer, and much milder, effects on the environment than do synthetic chemicals. What is more, they tend not to lead to the public health problems that chemicals are associated with. These two advantages of biological control agents, including bioherbicides, do not however mean that they need not also be subjected to careful tests for environmental and public health safety.

Bioherbicides may be compounds derived from microbes such as fungi, bacteria or protozoa; or phytotoxic plant residues, extracts or single compounds derived from other plant species.

A bioherbicide based on a fungus is called a mycoherbicide.

In the industry, bioherbicides and other biopesticides are often referred to as "naturals".

LUBILOSA

LUBILOSA was the name of a research programme that aimed at developing a biological alternative to the chemical control of locusts. This name is an acronym of the French title of the programme: Lutte Biologique contre les Locustes et les Sauteriaux (biological control of locusts and grasshoppers). During its 13 year life (November 1989 to December 2002), the programme identified an isolate of *Metarhizium* fungus virulent to locusts and went through all the necessary steps to develop the commercial biopesticide product 'Green Muscle' based on its spores.

Collaborators

The programme was conceived by Chris Prior and David Greathead of the International Institute of Biological Control (IIBC), a former research institute of CAB International that was based at Silwood Park in the UK at that time. IIBC requested and obtained the collaboration of the former Biological Control Centre for Africa in Cotonou, Benin, that was owned and managed by the International Institute of Tropical Agriculture (IITA). When IIBC asked for financial support from the Dutch Directoraat Generaal voor Internationale Samenwerking (DGIS: Directorate General of International Cooperation), DGIS contributed by paying for a Dutch locust expert based at the Département de Formation en Protection des Végétaux (DFPV: Department of Crop Protection Training) of the AGRHYMET Regional Centre (ARC) in Niamey, Niger, which is part of the Comité permanent Inter-états de Lutte contre la Sécheresse dans le Sahel (CILSS: Permanent Interstate Committee for Drought Control in the Sahel). Since DFPV was given its own budget within LUBILOSA, it became a full partner within the programme. For a number of years, LUBILOSA collaborated with GTZ's locust programme. Permanent collaboration was established with the crop protection agencies of Niger, Benin, Burkina Faso, Chad, Mali, Senegal and The Gambia.

Donors

The LUBILOSA programme was supported financially by the following donors:

- CIDA: Canadian International Development Agency
- DFID: Department for International Development, UK
- DGIS: Dutch Directorate General of International Cooperation
- SDC: Swiss Agency for Development and Cooperation
- USAID: United States Agency for International Development

Over the life of the programme, these donors contributed UK£10,200,000 (approximately US$17,000,000 at the time).

Four Phases

The LUBILOSA programme has been funded in phases of three or four years each:

- First phase 1990–1992
- Second phase 1993–1995
- Third phase 1996–1998
- Fourth phase 1999–2002

First Phase

The project was started late in 1989, with Chris Prior and David Greathead, obtaining funding and forming a team to develop a biological means of controlling locusts and grasshoppers. While examining the various options for biological control, it soon became apparent that oil formulations of the spores of certain fungi belonging to the form phylum Deuteromycota (Anamorphic fungi) offered the most promising option. Such fungi will grow on artificial substrates and so can be mass-produced relatively quickly in large quantities. Their spores are lipophilic and therefore suspend much more readily in oils than in water. As most locust control is carried out using oil formulations at ultra-low volume (ULV) rates of application, developing an oil formulation of fungal spores enabled operators to use the same equipment for applying a chemical pesticide and maximise efficacy.

Usually, fungi are most active under moist conditions and in early attempts to use them as control agents, they were therefore applied in water formulations. However, the results in practice were often disappointing and the approach was largely abandoned during most of the 20th century. When interest in microbial control was revived, water formulations of fungi started to be used successfully in greenhouses, but the results in open fields continued to disappoint. In the 1980s, Prior had previously discovered that some anamorphic fungi are more effective when applied in oil.

During the first phase of LUBILOSA, the technical feasibility of using such oil formulations against locusts was demonstrated in the laboratory, then field cage and "arena" trials. An extensive survey was launched in West Africa and the Arabian Peninsula to look for isolates virulent to locusts and grasshoppers, because few such isolates were available in public collections. The survey made use of a network of collaborators and came up with some 180 isolates, many of which belonged to *Metarhizium acridum*, some to other *Metarhizium* species and to *Beauveria bassiana* with a few records of *Syngliocladium acridiorum* (syn. *Sorosporella*). Laboratory screening of these isolates showed that *M. acridum* was easily the most virulent species under warm conditions (30°C), and confirmed the selection of isolate IMI 330189 for further development. Subsequent trials in cages and arenas confirmed that oil formulations were infective even under very dry conditions. IMI 330189 and similar isolates were assigned to *Metarhizium flavoviride* in early papers, then *Metarhizium anisopliae* var. *acridum*, but are now described as *Metarhizium acridum* (Driver & Milner) J.F. Bisch., Rehner & Humber.

Second Phase

During phase 2, the oil formulation was tested in the field and shown to be effective. Field trials proved to be difficult to carry out on such highly mobile insects as grasshoppers and locusts. With locusts, fixed plots, unless several square kilometres in size, cannot be used, so it is necessary to follow hopper bands. Although there were many locust infestations during this phase, of both desert locusts (*Schistocerca gregaria*) and brown locusts (*Locustana pardalina*), the chemical treatment teams were also very active and little headway was therefore made in developing the techniques necessary to measure the effect of *Metarhizium* on unrestrained locusts. However, it was possible to scale up treatments against the variegated grasshopper (*Zonocerus variegatus*) in the forest zone and against Sahelian grasshoppers, especially the Senegalese grasshopper (*Oedaleus senegalensis*).

Contacts were made with several internationally renowned biopesticide producers to ascertain whether they would be able and willing to reliably meet LUBILOSA's spore requirements during phase 3, but none were able to make that commitment. Accordingly, part of the IITA station in Cotonou was converted into a spore production unit and the human and technical capacity enhanced to meet the expected spore requirements. The facility proved an excellent research unit and made it possible to refine realistic technical specifications and to test production, contamination control, spore separation, drying and packaging.

Third Phase

Field Trials: The LUBILOSA team continued to look for opportunities to test *Metarhizium* against locusts and grasshoppers. Field trials were conducted against Senegalese grasshoppers, African rice grasshoppers (*Hieroglyphus daganensis*), variegated grasshoppers, Sahelian tree locusts (*Anacridium melanorhodon*), brown locusts and desert locusts. These trials demonstrated that *Metarhizium* at a dosage of 50 g/ha could reduce grasshopper populations and those of tree locusts by 80-90% within two to three weeks. Population level control was more difficult to achieve with other locusts, but a significant effect was demonstrated on hopper bands notwithstanding the great difficulties in tracking individual bands.

Some trials against Senegalese grasshoppers included the chemical insecticide fenitrothion for comparison. Because of its rather short persistence, this chemical turned out to be much less effective than commonly believed. It was highly effective in producing an immediate sharp reduction in population densities, but continued hatching and remigration into the treated plots caused these densities to start increasing again after less than a week, until they reached or exceeded the original levels after about two to three weeks. *Metarhizium*, on the other hand, reduced population levels at a slower rate, but maintained low levels for at least one and a half months. The half-life of its spores on the vegetation was estimated at more than 7 days under rainy season conditions in the Sahel and, of course, some conidia must survive between seasons.

Grasshopper Behaviour

In the course of these field trials, it became clear that locusts and grasshoppers are able to detect that they have become infected and alter their behaviour accordingly. In particular, they spend more time basking in the sun, even in the middle of the day. Uninfected grasshoppers also bask in the sun when their body temperature is below the preferred 38-40°C. However, detailed investigations showed that infected grasshoppers increase their body temperature to up to 4° higher. This effect has been dubbed "behavioural fever" and is thought to slow down the infection just like a fever in warm-blooded animals does. This basking behaviour poses a potential problem for the fungus. The latter's preferred temperature is 28-30°C and it stops growing above 35°. Measurements of body temperatures in the field have shown that the basking behaviour reduces the time available for growth by many hours per day. Under certain conditions, this enables the insects to lay at least one egg pod before dying. Interestingly, healthy locusts and grasshoppers, especially females, take some time after fledging to reach maturity and lay eggs, during which time they build up sufficient fat to produce up to three egg pods. However, infected ones usually dispense with this fattening period and go on to produce eggs immediately after fledging. This generally wears them out so much that they die after laying one pod. So even if *Metarhizium* does not kill before or shortly after fledging, it usually allows the formation of only a single egg pod and thereby reduces the fecundity of infected females.

(Eco)toxicology

During phase 3, the programme started investigating the environmental impact of the chosen *Metarhizium* isolate (IMI 330189). Standard toxicological tests carried out by certified laboratories showed that the strain had practically no effect on mammals, birds and fish (e.g. LD_{50} for rats >2000 mg active ingredient per kg of body weight). Other toxicological and ecotoxicological tests were performed by the LUBILOSA team. An important goal was to show that its isolate only infected locusts and grasshoppers. Infection tests were therefore performed on a large number of invertebrates. The great majority could not be infected even by high doses. However, some species, like termites, honey bees, certain parasitoid wasps and silk worms, did get infected under laboratory conditions. Further experiments were carried out under semi-field and full field conditions, which showed that only silk worms suffered significant infection under those conditions. The conclusion from all these tests was that the LUBILOSA strain of *Metarhizium* was incapable of infecting vertebrates and infected only a small number of insect species under artificial conditions. The available studies indicate that under natural conditions, the isolate only infects species belonging to the Acrididea and the domesticated silk worm *Bombyx mori*.

Mass Production

The spore production unit in Cotonou made it possible to optimise the mass production of spores of the LUBILOSA strain at an appropriate level of technology. The production process chosen was one with a liquid phase and a solid phase. During the liquid phase, the fungus biomass is bulked up in a solution of sugar and yeast and may produce submerged conidia. Subsequently, the resulting broth is used to inoculate partially boiled and sterilised rice. The fungal mycelium invades the rice and then starts to sporulate. At this point, rice and fungus are slowly dried before separating the spores from the rice. A novel machine was invented to achieve this in an efficient way, which was subsequently further developed into the 'MycoHarvester'. The study and optimisation of the mass production process led to the adoption of quality standards for the produced spores, which could be imposed on those who wanted to produce the spores commercially under licence. Various aspects of quality control are important, including: levels of contamination (especially the absence of human pathogens), virulence to target pests, particle size spectrum and, not least, viable spore count. Extensive research was carried out on optimising storage of spores, which should be dry (<5% moisture content) and ideally be maintained under cool conditions.

Commercialisation

Two commercial companies agreed to look into the feasibility and economic viability of producing *Metarhizium* for the locust control market. These were Biological Control Products (BCP) near Durban, South Africa, and Natural Plant Protection (NPP) of Noguères, France (part of Calliope S.A.). The former registered the product under the name Green Muscle (deposited by CAB International), first in South Africa and subsequently in other southern and eastern African countries. BCP is the leading producer of biopesticides in Africa. Its main market is South Africa but its products are increasingly being sold in other countries. The company continues its production and promotion of Green Muscle. It was taken over by Becker Underwood in 2010. NPP, however, only obtained a temporary sales permit for the CILSS countries in West Africa, but was unable to produce Green Muscle using its own production process.

Fourth Phase

The last phase of LUBILOSA was characterised by promotion activities and stewardship of the commercial producer. Launching the product on the market and informing potential clients was not sufficient. The market for locust control products was (and still is) peculiar and difficult to penetrate. Most users are government or intergovernmental agencies involved in locust and/or grasshopper control. Especially in the beginning, it proved difficult to convince the professionals in these agencies of the need for an alternative product and to convince them that the product was as effective as those they had been using so far. Moreover, the large pesticide companies offered all kinds of incentives to those in charge of procurement in order that they buy their products, a practice that was difficult to follow for a small company like BCP. Until the end of the programme and even beyond, there was a constant need to do more field trials or demonstrations. Towards the end, the Locust Group of the Food and Agriculture Organization of the United Nations (FAO) finally became interested and involved in some of the trials: especially against desert locusts.

Apart from the Sahelian zone, grasshoppers are generally not considered serious pests in Africa and are therefore rarely controlled with pesticides. For that reason, it was only in the Sahel that serious efforts were undertaken to promote Green Muscle for grasshopper control. In a number of countries, and especially Niger, cereal growers are organised in village brigades that are charged with controlling pests that surpass the capacity of individual growers. Green Muscle was given to some of these after training in its correct application. After the initial disappointment with its slow action, the villagers usually became satisfied with the final results. However, the price proved to be a serious obstacle. At that time, the cost of producing one kg of Green Muscle was about $200. At a dosage of 50 g/ha, that translated into a price of $10 per hectare. The price of conventional products was then approximately $5/ha. Though growers were prepared to pay a premium of up to $2 for the fact that Green Muscle was harmless to their health and that of their livestock, they were clearly not going to pay $10. Another problem was that no producer had yet been identified for the West African market and the availability of the product could therefore not be guaranteed.

At the end of the programme in December 2002, the LUBILOSA collaborators were faced with a situation where commercial uptake of Green Muscle was still non-existent, all quantities purchased so far exclusively for trial purposes. BCP was left alone to promote the product with little support from CABI and IITA, who had run out of funds for serious activities. Fortunately, the FAO had become interested, which made it possible to carry out a few more field trials.

Post-LUBILOSA

After the termination of the LUBILOSA programme, activities to promote the use of Green Muscle continued, including field trials against red locusts and desert locusts with support from the FAO. A new project in West Africa, PRéLISS (Regional Project for Integrated Grasshopper Control in the Sahel), aimed at developing an integrated pest management (IPM) strategy that included the use of Green Muscle. This project conducted field trials with reduced dosages of Green Muscle and proved that 25 g/ha was as effective as the registered dosage of 50 g/ha. It also tested mixtures with lambda-cyhalothrin (a pyrethroid insecticide) to try to solve the problem of the slow action of Green Muscle. It was demonstrated that application of both products together, each at a quarter of its registered dose, caused mortality almost as quick as lambda-cyhalothrin alone and maintained low population levels for weeks, and all this at a much lower cost.

After it had become clear that NPP could not produce Green Muscle, the search for a second producer continued. A number of candidate producers were approached, but none was ready to take up the challenge. This situation only changed when desert locusts started invading West African countries again in late 2003.

The president of Senegal, Abdoulaye Wade, and his wife Viviane were very concerned about the prospect of huge quantities of chemical insecticides being sprayed to contain the upsurge. The first lady instructed her technical advisor, Sébastien Couasnet, to find out whether an alternative existed. Apparently, the information about Green Muscle had remained stuck at the professional level of the crop protection agency of Senegal (and most likely other countries as well), notwithstanding all the efforts of LUBILOSA and PRéLISS to make the existence of the new product widely known.

Fortunately, Couasnet stumbled upon the LUBILOSA website, which was still hosted by IITA and maintained by the then project leader of PRéLISS, Christiaan Kooyman. The two met in Cotonou and immediately set about to design a production plant for Green Muscle. Couasnet was able to convince the first lady to organise the financing and construction of the plant started in 2005. It took a while to start up the production and get rid of all sources of contamination, but this was eventually achieved with the backstopping of IITA, CAB International and Roy Bateman of IPARC, who installed the latest version of the MycoHarvester v.3. The production licence was granted in late 2007. For the following four years, production and marketing were taken care of by the Fondation Agir pour l'Education et la Santé, of which Mme. Wade was president. Unfortunately, FAES did not last but went under because of internal disagreements, especially concerning the extent to which private capital should be brought in to perpetuate the production of Green Muscle and similar products. After the defeat of President Wade in the presidential elections of 2012, FAES was closed down. Production of fungi had already stopped 8 months earlier. Currently, the only remaining producer of Green Muscle is Becker Underwood South Africa, which was acquired by BASF SE in 2012.

Chitosan (Natural Biocontrol for Agricultural & Horticultural use)

Chitosan

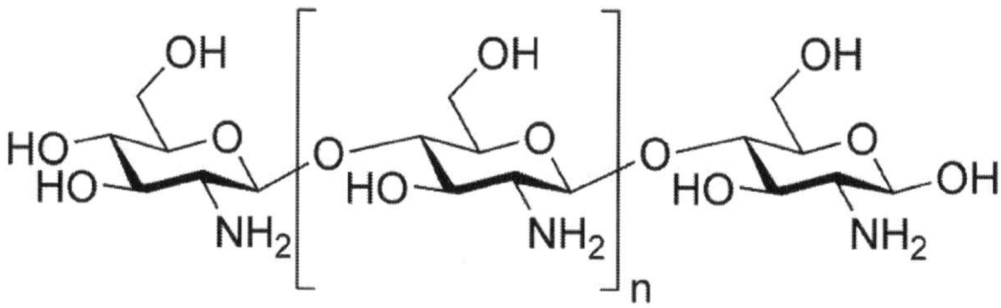

Figure: *Chitosan is a linear polysaccharide composed of randomly distributed β-(1-4)-linked D-glucosamine (deacetylated unit) and N-acetyl-D-glucosamine (acetylated unit). It is made by treating shrimp and other crustacean shells with the alkali sodium hydroxide.*

Chitosan has a number of commercial and possible biomedical uses. It can be used in agriculture as a seed treatment and biopesticide, helping plants to fight off fungal infections. In winemaking it can be used as a fining agent, also helping to prevent spoilage. In industry, it can be used in a self-healing polyurethane paint coating. In medicine, it may be useful in bandages to reduce bleeding and as an antibacterial agent; it can also be used to help deliver drugs through the skin.

More controversially, chitosan has been asserted to have use in limiting fat absorption, which would make it useful for dieting, but there is evidence against this.

Other uses of chitosan that have been researched include use as a soluble dietary fibre.

Manufacture and Properties

Chitosan is produced commercially by deacetylation of chitin, which is the structural element in the exoskeleton of crustaceans (such as crabs and shrimp) and cell walls of fungi. The degree of deacetylation (%DD) can be determined by NMR spectroscopy, and the %DD in commercial chitosans ranges from 60 to 100%. On average, the molecular weight of commercially produced chitosan is between 3800 and 20,000 Daltons. A common method for the synthesis of chitosan is the deacetylation of chitin using sodium hydroxide in excess as a reagent and water as a solvent. This reaction pathway, when allowed to go to completion (complete deacetylation) yields up to 98% product.

The amino group in chitosan has a pKa value of ~6.5, which leads to a protonation in acidic to neutral solution with a charge density dependent on pH and the %DA-value. This makes chitosan water soluble and a bioadhesive which readily binds to negatively charged surfaces such as mucosal membranes. Chitosan enhances the transport of polar drugs across epithelial surfaces, and is biocompatible and biodegradable. It is not approved by FDA for drug delivery though. Purified quantities of chitosans are available for biomedical applications.

Chitosan and its derivatives, such as trimethylchitosan (where the amino group has been trimethylated), have been used in nonviral gene delivery. Trimethylchitosan, or quaternised chitosan, has been shown to transfect breast cancer cells, with increased degree of trimethylation increasing the cytotoxicity; at approximately 50% trimethylation, the derivative is the most efficient at gene delivery. Oligomeric derivatives (3-6 kDa) are relatively nontoxic and have good gene delivery properties.

Usage

Agricultural and Horticultural Use

The agricultural and horticultural uses for chitosan, primarily for plant defence and yield increase, are based on how this glucosamine polymer influences the biochemistry and molecular biology of the plant cell. The cellular targets are the plasma membrane and nuclear chromatin. Subsequent changes occur in cell membranes, chromatin, DNA, calcium, MAP Kinase, oxidative burst, reactive oxygen species, callose pathogenesis-related (PR) genes and phytoalexins.

Natural Biocontrol and Elicitor

In agriculture, chitosan is used primarily as a natural seed treatment and plant growth enhancer, and as an ecologically friendly biopesticide substance that boosts the innate ability of plants to defend themselves against fungal infections. The natural biocontrol active ingredients, chitin/chitosan, are found in the shells of crustaceans, such as lobsters, crabs, and shrimp, and many other organisms, including insects and fungi. It is one of the most abundant biodegradable materials in the world.

Degraded molecules of chitin/chitosan exist in soil and water. Chitosan applications for plants and crops are regulated by the EPA, and the USDA National Organic Programme regulates its use on organic certified farms and crops. EPA-approved, biodegradable chitosan products are allowed for use outdoors and indoors on plants and crops grown commercially and by consumers.

The natural biocontrol ability of chitosan should not be confused with the effects of fertilizers or pesticides upon plants or the environment. Chitosan active biopesticides represent a new tier of cost-effective biological control of crops for agriculture and horticulture. The biocontrol mode of action of chitosan elicits natural innate defence responses within plant to resist insects, pathogens, and soil-borne diseases when applied to foliage or the soil. Chitosan increases photosynthesis, promotes and enhances plant growth, stimulates nutrient uptake, increases germination and sprouting, and boosts plant vigour. When used as seed treatment or seed coating on cotton, corn, seed potatoes, soybeans, sugar beets, tomatoes, wheat and many other seeds, it elicits an innate immunity response in developing roots which destroys parasitic cyst nematodes without harming beneficial nematodes and organisms.

Agricultural applications of chitosan can reduce environmental stress due to drought and soil deficiencies, strengthen seed vitality, improve stand quality, increase yields, and reduce fruit decay of vegetables, fruits and citrus crops. Horticultural application of chitosan increases blooms and extends the life of cut flowers and Christmas trees. The US Forest Service has conducted research on chitosan to control pathogens in pine trees and increase resin pitch outflow which resists pine beetle infestation.

Chitosan has a rich history of being researched for applications in agriculture and horticulture dating back to the 1980s. By 1989, chitosan salt solutions were applied to crops for improved freeze protection or to crop seed for seed priming. Shortly thereafter, chitosan salt received the first ever biopesticide label from the EPA, then followed by other intellectual property applications.

Chitosan has been used to protect plants in space, as well, exemplified by NASA's experiment to protect adzuki beans grown aboard the space shuttle and Mir space station in 1997. NASA results revealed chitosan induces increased growth (biomass) and pathogen resistance due to elevated levels of beta 1-3 glucanase enzymes within plant cells. NASA confirmed chitosan elicits the same effect in plants on earth.

Nontoxic, low molecular weight chitosan polymer solutions appear to be safe enough for broad-spectrum agricultural and horticultural uses. In 2008, the EPA approved natural broad-spectrum elicitor status for an ultralow molecular active ingredient of 0.25% chitosan.

A natural chitosan elicitor solution for agriculture and horticultural uses was granted an amended label for foliar and irrigation applications by the EPA in 2009. Given its low potential for toxicity and abundance in the natural environment, chitosan does not harm people, pets, wildlife, or the environment when used according to label directions. The US Forest Service tested chitosan as an ecofriendly biopesticide to prearm pine trees to defend themselves against mountain pine beetles.

Companion Planting

Companion planting in gardening and agriculture is the planting of different crops in proximity for pest control, pollination, providing habitat for beneficial creatures, maximizing use of space, and to otherwise increase crop productivity. Companion planting is a form of polyculture.

Companion planting is used by farmers and gardeners in both industrialized and developing countries for many reasons. Many of the modern principles of companion planting were present many centuries ago in cottage gardens in England and home gardens in Asia, and thousands of years ago in Mesoamerica.

In China, mosquito ferns (*Azolla* spp.) have been used for at least one thousand years as companion plants for rice crops. They host a cyanobacterium that fixes nitrogen from the atmosphere, and they block light from plants that would compete with the rice.

Companion planting was practised in various forms by the indigenous peoples of the Americas prior to the arrival of Europeans. These peoples domesticated squash 8,000 to 10,000 years ago, then maize, then common beans, forming the Three Sisters agricultural technique. The cornstalk served as a trellis for the beans to climb, and the beans fixed nitrogen, benefitting the maize.

Companion planting was widely promoted in the 1970s as part of the organic gardening movement. It was encouraged for pragmatic reasons, such as natural trellising, but mainly with the idea that different species of plant may thrive more when close together. It is also a technique frequently used in permaculture, together with mulching, polyculture, and changing of crops.

Examples of Companion Plants

Nasturtium (*Tropaeolum majus*) is a food plant of some caterpillars which feed primarily on members of the cabbage family (brassicas), and some gardeners claim that planting them around brassicas protects the food crops from damage, as eggs of the pests are preferentially laid on the nasturtium. This practice is called trap cropping.

The smell of the foliage of marigolds is claimed to deter aphids from feeding on neighbouring crops. Marigolds with simple flowers also attract nectar-feeding adult hoverflies, the larvae of which are predators of aphids.

Versions

There are a number of systems and ideas using companion planting.

Square foot gardening attempts to protect plants from many normal gardening problems by packing them as closely together as possible, which is facilitated by using companion plants, which can be closer together than normal.

Another system using companion planting is the forest garden, where companion plants are intermingled to create an actual ecosystem, emulating the interaction of up to seven levels of plants in a forest or woodland.

Organic gardening may make use of companion planting, since many synthetic means of fertilizing, weed reduction and pest control are forbidden.

Host-finding Disruption

Recent studies on host-plant finding have shown that flying pests are far less successful if their host-plants are surrounded by any other plant or even "decoy-plants" made of green plastic, cardboard, or any other green material.

The host-plant finding process occurs in phases:

- The first phase is stimulation by odours characteristic to the host-plant. This induces the insect to try to land on the plant it seeks. But insects avoid landing on brown (bare) soil. So if only the host-plant is present, the insects will quasi-systematically find it by simply landing on the only green thing around. This is called (from the point of view of the insect) "appropriate landing". When it does an "inappropriate landing", it flies off to any other nearby patch of green. It eventually leaves the area if there are too many 'inappropriate' landings.
- The second phase of host-plant finding is for the insect to make short flights from leaf to leaf to assess the plant's overall suitability. The number of leaf-to-leaf flights varies according to the insect species and to the host-plant stimulus received from each leaf. The insect must accumulate sufficient stimuli from the host-plant to lay eggs; so it must make a certain number of consecutive 'appropriate' landings. Hence if it makes an 'inappropriate landing', the assessment of that plant is negative, and the insect must start the process anew.

Thus it was shown that clover used as a ground cover had the same disruptive effect on eight pest species from four different insect orders. An experiment showed that 36% of cabbage root flies laid eggs beside cabbages growing in bare soil (which resulted in no crop), compared to only 7% beside cabbages growing in clover (which allowed a good crop). Simple decoys made of green cardboard also disrupted appropriate landings just as well as did the live ground cover.

Companion Plant Categories

The use of companion planting can be of benefit to the grower in a number of different ways, including:

- Hedged investment – the growing of different crops in the same space increases the odds of some yield being given, even if one crop fails.
- Increased level interaction – when crops are grown on different levels in the same space, such as providing ground cover or one crop working as a trellis for another, the overall yield of a plot may be increased.
- Protective shelter is when one type of plant may serve as a wind break or provide shade for another.

- Pest suppression – some companion plants may help prevent pest insects or pathogenic fungi from damaging the crop, through chemical means.
- Predator recruitment and positive hosting – The use of companion plants that produce copious nectar or pollen in a vegetable garden (insectary plants) may help encourage higher populations of beneficial insects that control pests, as some beneficial predatory insects only consume pests in their larval form and are nectar or pollen feeders in their adult form.
- Trap cropping – some companion plants are claimed to attract pests away from others.
- Pattern disruption – in a monoculture pests spread easily from one crop plant to the next, whereas such easy progress may be disrupted by surrounding companion plants of a different type.

Insectary Plants

Insectary plants are those that attract insects. As such, beneficial insectary plants are intentionally introduced into an ecosystem to increase pollen resources and nectar resources required by the natural enemies of harmful or unwanted insect pests. Beyond an effective natural control of pests, the *friendly insects* also assist in pollination.

The "friendly insects" include ladybeetles, bees, ground beetles, hoverflies, and parasitic wasps. Other animals that are frequently considered beneficial include lizards, spiders, toads, and humming birds. Beneficial insects are as much as ten times more abundant in the insectary plantings area. Mortality of scale insects (caused by natural enemies) can be double with insectary plantings. In addition, a diversity of insectary plants can increase the population of beneficial insects such that these levels can be sustained even when the insectary plants are removed or die off.

For maximum benefit in the garden, insectary plants can be grown alongside desired garden plants that do not have this benefit. The insects attracted to the insectary plants will also help the other nearby garden plants.

Many members of the Apiaceae (formerly known as Umbelliferae) family are excellent insectary plants. Fennel, angelica, coriander (cilantro), dill, and wild carrot all provide in great number the tiny flowers required by parasitic wasps. Various clovers, yarrow, and rue also attract parasitic and predatory insects. Low-growing plants, such as thyme, rosemary, or mint, provide shelter for ground beetles and other beneficial insects. Composite flowers (daisy and chamomile) and mints (spearmint, peppermint, or catnip) will attract predatory wasps, hoverflies, and robber flies. The wasps will catch caterpillars and grubs to feed their young, while the predatory and parasitic flies attack many kinds of insects, including leafhoppers and caterpillars.

Other insectary plants include: mustard (Brassica juncea), phacelia (Phacelia tanacetifolia), buckwheat (Fagopyrum esceulentum), marigold (Tagetes patula), elderberry (sambucus mexicana), and Korean licorice mint (Agastache rugosa), Rubus fruticosa, Convolvulus tricolour, Crataegus monogyna, Anthriscus sylvestris, Chrysanthemum segetum, Scrophularia, Rosa canina, Hedera helix, Centaurea cyanus, Phacelia tanacetifolia, Escholzia californica, Prunus spinosa, Lobularia maritima

Inundative Application

Inundative biological control refers to the release of overwhelming numbers of a mass-produced biological control agent in the expectation of achieving a rapid reduction of a pest population without necessarily achieving continuing impact.

Japanese Beetle

The beetle species *Popillia japonica* is commonly known as the Japanese beetle. It is about 15 millimetres (0.6 in) long and 10 millimetres (0.4 in) wide, with iridescent copper-coloured elytra and green thorax and head. It is not very destructive in Japan, where it is controlled by natural predators, but in America it is a serious pest of about 200 species of plants, including rose bushes, grapes, hops, canna, crape myrtles, birch trees, linden trees and others.

It is a clumsy flier, dropping several centimetres when it hits a wall. Japanese beetle traps therefore consist of a pair of crossed walls with a bag or plastic container underneath, and are baited with floral scent, pheromone, or both.

However, studies conducted at the University of Kentucky and Eastern Illinois University suggest beetles attracted to traps frequently do not end up in the traps, but alight on plants in the vicinity, thus causing more damage along the flight path of the beetles and near the trap than may have occurred if the trap were not present.

These insects damage plants by skeletonizing the foliage, that is, consuming only the leaf material between the veins, and may also feed on fruit on the plants if present.

Control

During the larval stage, the Japanese beetle lives in lawns and other grasslands, where it eats the roots of grasses. During that stage, it is susceptible to a fatal disease called milky spore disease, caused by a bacterium called milky spore, *Paenibacillus* (formerly *Bacillus*) *popilliae*. The USDA developed this biological control and it is commercially available in powder form for application to lawn areas. Standard applications (low density across a broad area) take from one to five years to establish maximal protection against larval survival (depending on climate), expanding through the soil through repeated rounds of infection.

On field crops such as squash, floating row covers can be used to exclude the beetles, however this may necessitate hand pollination of flowers. Kaolin sprays can also be used as barriers.

Research performed by many US extension service branches has shown pheromone traps attract more beetles than they catch. Traps are most effective when spread out over an entire community, and downwind and at the borders (i.e., as far away as possible, particularly upwind), of managed property containing plants being protected. Natural repellents include catnip, chives, garlic, and tansy, as well as the remains of dead beetles, but these methods have limited effectiveness. Additionally, when present in small numbers, the beetles may be manually controlled using a soap-water spray mixture, shaking a plant in the morning hours and disposing of the fallen beetles, or simply picking them off attractions such as rose flowers, since the presence of beetles attracts more beetles to that plant.

Natural predators of the beetle are the anchor bug and the blue-winged wasp.

Mating Disruption

Mating disruption (MD) is a pest management technique designed to control certain insect infestations. Specifically, mating disruption involves the use of synthesized sex pheromones to disrupt the reproductive cycle of insects.

In many insect species of interest to agriculture, such as those in the order Lepidoptera, females emit an airborne trail of a specific chemical blend constituting that species' sex pheromone. This aerial trail is referred to as a pheromone plume.

Males of that species use the information contained in the pheromone plume to locate the emitting female (known as a "calling" female). Mating disruption exploits the male insects' natural response to follow the plume by introducing a synthetic pheromone into the insects' habitat. The synthetic pheromone is a volatile organic chemical designed to mimic the species-specific sex pheromone produced by the female insect. The general effect of mating disruption is to confuse the male insects by masking the natural pheromone plumes, causing the males to follow "false pheromone trails" at the expense of finding mates, and affecting the males' ability to respond to "calling" females.

Consequently, the male population experiences a reduced probability of successfully locating and mating with females, which leads to the eventual cessation of breeding and collapse of the insect infestation. The California Department of Pesticide Regulation, the California Department of Food and Agriculture, and the United States Environmental Protection Agency consider mating disruption to be among the most environmentally friendly treatments used to eradicate pest infestations. Mating disruption works best if large areas are treated with pheromones. Ten acres is a good minimum size for a successful MD programme, but larger areas are preferable

Advantages of Mating Disruption

Pheromone programmes are most effective when controlling low to moderate pest population densities. MD has also been identified as a pest control method in which the insect does not become resistant. The scientific community, together with governmental agencies throughout the world, understands the benefits of mating disruption using species-specific sex pheromones, and consider sex-pheromone-based insect control programmes among the most environmentally friendly treatments to be used to manage and control insect pest populations.

Insect pheromone has been successfully used as an effective tool to slow the spread and to eradicate pests from very large areas in the US; for example to control the Gypsy moth (Lymantria dispar), a devastating forestry pest, and to eradicate the boll weevil and pink bollworm, two of the most damaging pest of cotton. Conventional pesticide based control methods, kill insects directly, whereas mating disruption confuses male insects from accurately locating a mating partner, leading to the eventual collapse of the mating cycle. Mating disruption, due to the specificity of the sex pheromone of the insect species, has the benefit of only affecting the males of that species, while leaving other non target species unaffected. This allows for very targeted pest management, promoting the suppression of a single pest species, leaving the populations of beneficial insects (pollinators and natural enemies) intact. Mating disruption, like most pest management strategies, is a useful technique, but should not be considered a stand-alone treatment programme for it targets only a single species in plant production systems that usually have several pests of concern. Mating disruption is a valuable tool that should be used in Integrated Pest Management(IPM) programmes.

Pheromone programmes have been used for several decades around the globe and to date (2009) there is no documented public health evidence to suggest that agricultural use of synthetic pheromones is harmful to humans or to any other non-target species. However, continuing research is being conducted.

Disadvantages of Mating Disruption

Over the decades that pheromone pest programmes have been used several disadvantages have been argued when compared to the use of conventional pesticides. Most pheromones target a single species, so a specific mating disruption formulation controls only the species that uses that pheromone blend; whereas pesticides usually kill indiscriminately a plethora of species, including multiple species with a single application. Some synthetic pheromones have high developmental and production costs, causing the mating disruption technique to be too costly to be adopted by conventional commercial growers. Furthermore most commercial pheromone mating disruption formulations must be applied by hand, which can be an expensive and time consuming. Novel pheromone formulations recently developed to be mechanically applied provide long lasting mating disruption effects (e.g., depending on the target pest a single application of SPLAT controls the target pest for a complete reproductive cycle, or for the entire season

Methods of Dispersal

Microencapsulated Pheromones

Microencapsulated pheromones (MECs) are small droplets of pheromone enclosed within polymer capsules. The capsules control the release rate of the pheromone into the surrounding environment. The capsules are small enough to be applied in the same method as used to spray insecticides. The effective field longevity of the microencapsulated pheromone formulations ranges from a few days to slightly more than a week, depending on climatic conditions, capsule size and chemical properties. Microcapsules in the pheromone formulations are usually kept above a prescribed diametre to avoid the risk of inhalation by humans.

Hand Applied Dispensers

- Hollow tube dispensers are plastic twist-tie type dispensers, plastic hollow fibres or plastic hollow microfibres fibres, filled with synthetic sex pheromone and placed throughout the area to be protected.
- Pheromone Baits and Stations are a stationary, attract and kill type of dispensers. Some are relatively large platform, containing a pheromone lure inside a glue board that ensnares the attracted insect. Other pheromone bait stations contain a pheromone lure in conjunction with a surface containing a dose of insecticide that reduces the attracted insect's fitness, thus reducing its ability to mate and reproduce.
- High-emission dispensers There are several very high dose pheromone dispensers, some do it passively, like pheromone sachets and large dollops of SPLAT pheromone formulations, others do it be actively releasing bursts of sex pheromone at timed intervals.

Monolithic Flowable Dispensers

A new, effective and economical concept in pheromone delivery using a flowable formulation to create long lasting monolithic pheromone dispensers has been brought to the market in the past decade. These novel SPLAT pheromone mating disruption formulations can provide effective season long suppression effect (e.g., depending on the target pest a single application of SPLAT controls the target pest for a complete reproductive cycle, or for the entire season) and can be manually or mechanically applied. Although mechanical dispersal techniques require specialized off-the-shelf application technology and/or equipment, once the application system is made to work it allows protection of extensive areas using pheromones, one of the most benign and effective pest management techniques available today. A benefit of SPLAT is that the dollop anchors where it lands, avoiding unwanted drift of the formulation once applied in the field, and, depending on the mode of application, the cured dollops are retrievable.

Aerial Dispersal

In November 2007, a controversial aerial approach was used to spray microencapsulated LBAM pheromone in urban and rural areas of the counties of Santa Cruz and Monterey California to combat the invasive light brown apple moth. Usually the effect of disruption of orientation of the male moths to females (or monitoring pheromone traps) can be detected by the reduction in moth capture in monitoring pheromone traps.

The government campaign using areawide aerial microencapsulated pheromone applications failed to show any sign of mating disruption on the light brown apple moth populations in the treated area. It was found that the first aerial campaign was performed using an incomplete (the wrong) pheromone blend of the light brown apple moth (the wrong blend decreased tremendously the likelihood of success of the mating disruption programme), and the LBAM microencapsulated formulation was untested, and finally, microencapsule formulations are notoriously known for their short field life, weak and erratic performance. Furthermore it is possible that the LBAM microencapsulated formulation used in the government campaign was unfit for aerial delivery in urban areas; although pheromone is safe, the formulation used had microcapsules of very small diametre which made it into a possible inhalation hazard that seems to be linked to an increase in allergenic reactions of the population in the target area. This set of LBAM mating disruption aerial applications done by the government has created tremendous dissent of the public in general as well as of several sectors of the scientific community. Now, several years later, the affected communities as well as the nascent US pheromone industry (which provides safer, yet very effective, alternatives to the use of conventional pesticides) are still suffering the ripple effects of these disastrous Bay Area LBAM eradication campaigns.

But there are numerous, successful pest suppression programmes that rely on aerial dispersal of pheromone mating disruptants. One of the largest pheromone mating disruption programmes in the globe is the Gypsy Moth Slow the Spread. Gypsy Moth Slow the Spread has been implemented across the 1,200-mile (1,900 km) gypsy moth frontier from Wisconsin to North Carolina.

The programme area is located ahead of the advancing front of the gypsy moth population. The STS programme focuses on early detection and suppression of the low–level populations along this advancing front, disrupting the natural progress of population buildup and spread. Every year hundreds of thousands of acres are aerially sprayed with two pheromone Gypsy moth pheromone mating disruption formulations, Flakes and SPLAT. A single mating disruption formulation application promotes season-long suppression of gypsy moth in the treated areas. With a crew of 8 people it was possible to aerially treat with SPLAT GM over 20,000 acres (81 km^2) of forest in a single day. The consortium of Federal and State participants have been able to do the following:

- decrease the new territory invaded by the gypsy moth each year from 15,600 square miles (40,000 km^2) to 6,000 square miles (16,000 km^2);
- protect forests, forest–based industries, urban and rural parks, and private property; and
- avoid at least $22 million per year in damage and management costs.

It seems that the tremendous success of the Gypsy Moth Slow the Spread programme is related to extremely well planned campaigns, which involves communication, transparency and clarity of objectives: in advance to an application STS holds meetings that include the area population in general, concerned citizens, public officials, scientists and technical personnel to discuss strategies of management of Gypsy moth in the areas of concern. There is a movement requesting that new government invasive species eradication campaigns model their pest suppression actions on the existing successful suppression programmes like GM STS, and embrace a more effective policy of communication, transparency and clarity of objectives.

With the involvement and education of the public, areawide eradication campaigns will be better planned and more able to deliver decisive end effective pest eradication actions.

Nematophagous Fungus

Nematophagous fungi are carnivorous fungi specialized in trapping and digesting nematodes. Around 160 species are known. There exist both species that live inside the nematodes from the beginning and others that catch them mostly with glue traps or in rings, some of which constrict on contact.

Some species possess both types of traps. Another technique is to stun the nematodes using toxins, which is a method employed by *Coprinus comatus*, *Stropharia rugosoannulata*, and the family Pleurotaceae.

Nematophagous fungus can be useful in controlling nematodes. *Paecilomyces* for example can be used as a bio-nematicide.

Nematicide

A nematicide is a type of chemical pesticide used to kill plant-parasitic nematodes. Nematicides have tended to be broad-spectrum toxicants possessing high volatility or other properties promoting migration through the soil. Temik (Aldicarb), a carbamate insecticide marketed by Bayer CropScience, is an example of a commonly used commercial nematicide. It is important in potato production, where it has been used for control of soil-borne nematodes.

Aldicarb is a cholinesterase inhibitor, which prevents the breakdown of acetylcholine in the synapse. In case of severe poisoning, the victim dies of respiratory failure. It is no longer authorised for use in the EU and, in August, 2010, Bayer CropScience announced that it plans to discontinue Aldicarb by 2014. Human health safety and environmental concerns have resulted in the widespread deregistration of several other agronomically important nematicides.

Prior to 1985, the persistent halocarbon DBCP was a widely used nematicide and soil fumigant.

However, it was banned from use after being linked to sterility among male workers; the Dow Chemical company was subsequently found liable for more than $600 million in damages.

Natural Nematicides

Several natural nematicides are known. An environmentally benign garlic-derived polysulfide product is approved for use in the European Union (under Annex 1 of 91/414) and the UK as a nematicide. Another common natural nematicide is obtained from neem cake, the residue obtained after cold-pressing the fruit and kernels of the neem tree. Known by several names in the world, the tree was first cultivated in India in ancient times and is now widely distributed throughout the world. The root exudate of marigold (*Tagetes*) is also found to have nematicidal action. Nematophagous fungi, a type of carnivorous fungi, can be useful in controlling nematodes, *Paecilomyces* being one example.

Besides chemicals, soil steaming can be used in order to kill nematodes. Superheated steam is induced into the soil, which causes almost all organic material to deteriorate.

Chapter 2

Sustainable Farming

Sustainable agriculture is the act of farming using principles of ecology, the study of relationships between organisms and their environment. The phrase was reportedly coined by Australian agricultural scientist Gordon McClymont. It has been defined as "an integrated system of plant and animal production practices having a site-specific application that will last over the long term" For Example:

- Satisfy human food and fibre needs
- Enhance environmental quality and the natural resource base upon which the agricultural economy depends
- Make the most efficient use of non-renewable resources and on-farm resources and integrate, where appropriate, natural biological cycles and controls
- Sustain the economic viability of farm operations
- Enhance the quality of life for farmers and society as a whole

Farming and Natural Resources

Sustainable agriculture can be understood as an ecosystem approach to agriculture. Practices that can cause long-term damage to soil include excessive tilling of the soil(leading to erosion) and irrigation without adequate drainage (leading to salinization). Long-term experiments have provided some of the best data on how various practices affect soil properties essential to sustainability. In the United States a federal agency, USDA-Natural Resources Conservation Service, specializes in providing technical and financial assistance for those interested in pursuing natural resource conservation and production agriculture as compatible goals.

The most important factors for an individual site are sun, air, soil, nutrients, and water. Of the five, water and soil quality and quantity are most amenable to human intervention through time and labour.

Although air and sunlight are available everywhere on Earth, crops also depend on soil nutrients and the availability of water. When farmers grow and harvest crops, they remove some of these nutrients from the soil. Without replenishment, land suffers from nutrient depletion and becomes either unusable or suffers from reduced yields. Sustainable agriculture depends on replenishing the soil while minimizing the use or need of non-renewable resources, such as natural gas (used in converting atmospheric nitrogen into synthetic fertilizer), or mineral ores (e.g., phosphate). Possible sources of nitrogen that would, in principle, be available indefinitely, include:

1. recycling crop waste and livestock or treated human manure
2. growing legume crops and forages such as peanuts or alfalfa that form symbioses with nitrogen-fixing bacteria called rhizobia
3. industrial production of nitrogen by the Haber Process uses hydrogen, which is currently derived from natural gas, (but this hydrogen could instead be made by electrolysis of water using electricity (perhaps from solar cells or windmills)) or
4. genetically engineering (non-legume) crops to form nitrogen-fixing symbioses or fix nitrogen without microbial symbionts.

The last option was proposed in the 1970s, but is only recently becoming feasible. Sustainable options for replacing other nutrient inputs (phosphorus, potassium, etc.) are more limited. More realistic, and often overlooked, options include long-term crop rotations, returning to natural cycles that annually flood cultivated lands (returning lost nutrients indefinitely) such as the Flooding of the Nile, the long-term use of biochar, and use of crop and livestock landraces that are adapted to less than ideal conditions such as pests, drought, or lack of nutrients.

Crops that require high levels of soil nutrients can be cultivated in a more sustainable manner if certain fertilizer management practices are adhered to.

Water

In some areas sufficient rainfall is available for crop growth, but many other areas require irrigation. For irrigation systems to be sustainable, they require proper management (to avoid salinization) and must not use more water from their source than is naturally replenishable. Otherwise, the water source effectively becomes a non-renewable resource. Improvements in water well drilling technology and submersible pumps, combined with the development of drip irrigation and low pressure pivots, have made it possible to regularly achieve high crop yields in areas where reliance on rainfall alone had previously made successful agriculture unpredictable. However, this progress has come at a price. In many areas, such as the Ogallala Aquifer, the water is being used faster than it can be recharged.

Several steps must be taken to develop drought-resistant farming systems even in "normal" years with average rainfall. These measures include both policy and management actions:

1) improving water conservation and storage measures,
2) providing incentives for selection of drought-tolerant crop species,
3) using reduced-volume irrigation systems,
4) managing crops to reduce water loss, or
5) not planting crops at all.

Indicators for Sustainable Water Resource Development

Internal Renewable Water Resources: This is the average annual flow of rivers and groundwater generated from endogenous precipitation, after ensuring that there is no double counting. It represents the maximum amount of water resource produced within the boundaries of a country. This value, which is expressed as an average on a yearly basis, is invariant in time (except in the case of proved climate change). The indicator can be expressed in three different units: in absolute terms (km3/yr), in mm/yr (it is a measure of the humidity of the country), and as a function of population (m3/ person per yr).

Global Renewable Water Resources: This is the sum of internal renewable water resources and incoming flow originating outside the country. Unlike internal resources, this value can vary with time if upstream development reduces water availability at the border. Treaties ensuring a specific flow to be reserved from upstream to downstream countries may be taken into account in the computation of global water resources in both countries.

Dependency Ratio: This is the proportion of the global renewable water resources originating outside the country, expressed in percentage. It is an expression of the level to which the water resources of a country depend on neighbouring countries.

Water Withdrawal: In view of the limitations described above, only gross water withdrawal can be computed systematically on a country basis as a measure of water use. Absolute or per-person value of yearly water withdrawal gives a measure of the importance of water in the country's economy. When expressed in percentage of water resources, it shows the degree of pressure on water resources. A rough estimate shows that if water withdrawal exceeds a quarter of global renewable water resources of a country, water can be considered a limiting factor to development and, reciprocally, the pressure on water resources can have a direct impact on all sectors, from agriculture to environment and fisheries.

Soil

Figure: *Walls built to avoid water run-off*

Soil erosion is fast becoming one of the worlds greatest problems. It is estimated that "more than a thousand million tonnes of southern Africa's soil are eroded every year. Experts predict that crop yields will be halved within thirty to fifty years if erosion continues at present rates." Soil erosion is not unique to Africa but is occurring worldwide. The phenomenon is being called *Peak Soil* as present large scale factory farming techniques are jeopardizing humanity's ability to grow food in the present and in the future. Without efforts to improve soil management practices, the availability of arable soil will become increasingly problematic.

Some Soil Management techniques

1. No-till farming
2. Keyline design
3. Growing wind breaks to hold the soil
4. Incorporating organic matter back into fields

5. Stop using chemical fertilizers (which contain salt)
6. Protecting soil from water run off(soil erosion)

Phosphate

Phosphate is a primary component in the chemical fertilizer which is applied in modern agricultural production. However, scientists estimate that rock phosphate reserves will be depleted in 50–100 years and that Peak phosphorus will occur in about 2030. The phenomenon of Peak phosphorus is expected to increase food prices as fertilizer costs increase as rock phosphate reserves become more difficult to extract. In the long term, phosphate will therefore have to be recovered and recycled from human and animal waste in order to maintain food production.

Land

As the global population increases and demand for food increases, there is pressure on land resources. Land can also be considered a finite resource on Earth. Expansion of agricultural land has an impact on biodiversity and contributes to deforestation. The Food and Agriculture Organisation of the United Nations estimates that in coming decades, cropland will continue to be lost to industrial and urban development, along with reclamation of wetlands, and conversion of forest to cultivation, resulting in the loss of biodiversity and increased soil erosion.

Energy for Agriculture

Energy is used all the way down the food chain from farm to fork. In industrial agriculture, energy is used in on-farm mechanisation, food processing, storage, and transportation processes. It has therefore been found that energy prices are closely linked to food prices. Oil is also used as an input in agricultural chemicals. Higher prices of non-renewable energy resources are projected by the International Energy Agency. Increased energy prices as a result of fossil fuel resources being depleted may therefore impact negatively on the global food security unless action is taken to 'decouple' fossil fuel energy from food production, with a move towards 'Energy-Smart' agricultural systems.

Economics

Socioeconomic aspects of sustainability are also partly understood. Regarding less concentrated farming, the best known analysis is Netting's study on smallholder systems through history. The Oxford Sustainable Group defines sustainability in this context in a much broader form, considering effect on all stakeholders in a 360 degree approach Given the finite supply of natural resources at any specific cost and location, agriculture that is inefficient or damaging to needed resources may eventually exhaust the available resources or the ability to afford and acquire them. It may also generate negative externality, such as pollution as well as financial and production costs.

The way that crops are sold must be accounted for in the sustainability equation. Food sold locally does not require additional energy for transportation (including consumers). Food sold at a remote location, whether at a farmers' market or the supermarket, incurs a different set of energy cost for materials, labour, and transport.

Methods

What grows where and how it is grown are a matter of choice. Two of the many possible practices of sustainable agriculture are crop rotation and soil amendment, both designed to ensure that crops being cultivated can obtain the necessary nutrients for healthy growth. Soil amendments would include using locally available compost from community recycling centres. These community recycling centres help produce the compost needed by the local organic farms.

Many scientists, farmers, and businesses have debated how to make agriculture sustainable. Using community recycling from yard and kitchen waste utilizes a local area's commonly available resources. These resources in the past were thrown away into large waste disposal sites, are now used to produce low cost organic compost for organic farming. Other practices includes growing a diverse number of perennial crops in a single field, each of which would grow in separate season so as not to compete with each other for natural resources.

This system would result in increased resistance to diseases and decreased effects of erosion and loss of nutrients in soil. Nitrogen fixation from legumes, for example, used in conjunction with plants that rely on nitrate from soil for growth, helps to allow the land to be reused annually. Legumes will grow for a season and replenish the soil with ammonium and nitrate, and the next season other plants can be seeded and grown in the field in preparation for harvest.

Monoculture, a method of growing only one crop at a time in a given field, is a very widespread practice, but there are questions about its sustainability, especially if the same crop is grown every year. Today it is realized to get around this problem local cities and farms can work together to produce the needed compost for the farmers around them. This combined with growing a mixture of crops (polyculture) sometimes reduces disease or pest problems but polyculture has rarely, if ever, been compared to the more widespread practice of growing different crops in successive years (crop rotation) with the same overall crop diversity. Cropping systems that include a variety of crops (polyculture and/or rotation) may also replenish nitrogen (if legumes are included) and may also use resources such as sunlight, water, or nutrients more efficiently.

Replacing a natural ecosystem with a few specifically chosen plant varieties reduces the genetic diversity found in wildlife and makes the organisms susceptible to widespread disease. The Great Irish Famine (1845–1849) is a well-known example of the dangers of monoculture. In practice, there is no single approach to sustainable agriculture, as the precise goals and methods must be adapted to each individual case. There may be some techniques of farming that are inherently in conflict with the concept of sustainability, but there is widespread misunderstanding on impacts of some practices.

Today the growth of local farmers' markets offer small farms the ability to sell the products that they have grown back to the cities that they got the recycled compost from. By using local recycling this will help move people away from the slash-and-burn techniques that are the characteristic feature of shifting cultivators are often cited as inherently destructive, yet slash-and-burn cultivation has been practised in the Amazon for at least 6000 years; serious deforestation did not begin until the 1970s, largely as the result of Brazilian government programmes and policies. To note that it may not have been slash-and-burn so much as slash-and-char, which with the addition of organic matter produces terra preta, one of the richest soils on Earth and the only one that regenerates itself.

There are also many ways to practice sustainable animal husbandry. Some of the key tools to grazing management include fencing off the grazing area into smaller areas called paddocks, lowering stock density, and moving the stock between paddocks frequently.

Several attempts have been made to produce an artificial meat, using isolated tissues to produce it in vitro; Jason Matheny's work on this topic, which in the New Harvest project, is one of the most commented.

Soil Treatment

Soil steaming can be used as an ecological alternative to chemicals for soil sterilization. Different methods are available to induce steam into the soil in order to kill pests and increase soil health. Community and farm composting of kitchen, yard, and farm organic waste can provide most if not all the required needs of local farms. This composting could potentially be a reliable source of energy.

Off-farm Impacts

A farm that is able to "produce perpetually", yet has negative effects on environmental quality elsewhere is not sustainable agriculture. An example of a case in which a global view may be warranted is over-application of synthetic fertilizer or animal manures, which can improve productivity of a farm but can pollute nearby rivers and coastal waters (eutrophication). The other extreme can also be undesirable, as the problem of low crop yields due to exhaustion of nutrients in the soil has been related to rainforest destruction, as in the case of slash and burn farming for livestock feed.In Asia, specific land for sustainable farming is about 12.5 acres which includes land for animal fodder, cereals productions lands for some cash crops and even recycling of related food crops.In some cases even a small unit of aquaculture is also included in this number (AARI-1996)

Sustainability affects overall production, which must increase to meet the increasing food and fibre requirements as the world's human population expands to a projected 9.3 billion people by 2050. Increased production may come from creating new farmland, which may ameliorate carbon dioxide emissions if done through reclamation of desert as in Israel and Palestine, or may worsen emissions if done through slash and burn farming, as in Brazil.

International Policy

Sustainable agriculture has become a topic of interest in the international policy arena, especially with regards to its potential to reduce the risks associated with a changing climate and growing human population.

The Commission on Sustainable Agriculture and Climate Change, as part of its recommendations for policy makers on achieving food security in the face of climate change, urged that sustainable agriculture must be integrated into national and international policy. The Commission stressed that increasing weather variability and climate shocks will negatively affect agricultural yields, necessitating early action to drive change in agricultural production systems towards increasing resilience. It also called for dramatically increased investments in sustainable agriculture in the next decade, including in national research and development budgets, land rehabilitation, economic incentives, and infrastructure improvement.

Urban Planning

There has been considerable debate about which form of human residential habitat may be a better social form for sustainable agriculture.

Many environmentalists advocate urban developments with high population density as a way of preserving agricultural land and maximizing energy efficiency. However, others have theorized that sustainable ecocities, or ecovillages which combine habitation and farming with close proximity between producers and consumers, may provide greater sustainability. The use of available city space (e.g., rooftop gardens, community gardens, garden sharing, and other forms of urban agriculture) for cooperative food production is another way to achieve greater sustainability.

One of the latest ideas in achieving sustainable agriculture involves shifting the production of food plants from major factory farming operations to large, urban, technical facilities called vertical farms. The advantages of vertical farming include year-round production, isolation from pests and diseases, controllable resource recycling, and on-site production that reduces transportation costs. While a vertical farm has yet to become a reality, the idea is gaining momentum among those who believe that current sustainable farming methods will be insufficient to provide for a growing global population.

Sustainable Gardening

Sustainable gardening (which is taken here to include sustainable landscapes, sustainable landscape design, sustainable landscape architecture and sustainable sites) comprises a disparate group of horticultural interests that share, to a greater or lesser extent, the aims and objectives associated with the international post-1980s sustainable development and sustainability programmes developed to address the fact that humans are now using natural biophysical resources faster than they can be replenished by nature. Included within this compass are those home gardeners, and members of the landscape and nursery industries and municipal authorities, that integrate environmental, social and economic factors in an attempt to create a more sustainable future.

Organic gardening and the use of native plants are integral to sustainable gardening.

After the establishment of sustainable agriculture in the early 1980s it was some time before the emergence of Sustainable

Horticulture (as sustainable *production* horticulture) at the International Society of Horticultural Science's First International Symposium on Sustainability in Horticulture held at the International Horticultural Congress in Toronto in 2002. This symposium produced "conclusions ... on Sustainability in Horticulture and a Declaration for the 21st Century". The principles and objectives outlined at this conference were discussed in more practical terms at the following conference at Seoul in 2006.

Many of the eco-friendly principles and ideas espoused by sustainable gardens, landscapes and sites perpetuate sustainable practices established as a reaction to resource-intensive industrial agriculture. These practices were established as movements for self-sufficiency and small-scale farming based on a holistic systems approach and ecological principles. Included here would be: biodynamic agriculture, no-till farming, agroecology, Fukuoka farming, forest gardening, organic gardening and others. On a larger scale there is the more recent "whole farm planning" which was established in 1995, and ecoagriculture established in 2000, and other variants of sustainable agricultural systems. Perhaps the most influential of these approaches is permaculture, established by Australians Bill Mollison and David Holmgren as both a design system and a loosely defined philosophy or lifestyle ethic. Permaculture shares many principles and practices of the above but not the broad philosophical base as indicated by the title of the 2002 publication *Permaculture, principles and pathways beyond sustainability*. The application of sustainability principles to the horticultural sphere has now becoming broadly accepted in commerce and academia.

Definition

The American Sustainable Sites Initiative is an interdisciplinary approach used by the American Society of Landscape Architects, the Lady Bird Johnson Wildflower Centre and the United States Botanic Garden to create voluntary national guidelines and performance benchmarks for sustainable land design, construction and maintenance practices: it was founded in 2005.

Principles and Concepts

Managing global biophysical cycles and ecosystem services for the benefit of humans, other organisms and future generations has now become a global human responsibility. The method of applying sustainability to gardens, landscapes and sites is still under development and varies somewhat according to the context under consideration. However, there are a number of basic and common underlying biological and operational principles and practices in the sustainable sites literature.

Biological Principles

Sustainable management of man-made landscapes emulates the natural processes that sustain the biosphere and its ecosystems. First and foremost is the harnessing the energy of the Sun and the cycling of materials thereby minimising waste and energy use. Running within, and dependent on, the natural economy there is the production and consumption of goods and services in the "human economy" which has now significantly altered, in a detrimental way, natural biogeochemical cycles (notable here are the water cycle, carbon cycle and nitrogen cycle so sustainable practices maximise support for ecosystem services.

Native Plants

The use of native plants in a garden or landscape can both preserve and protect natural ecosystems, and reduce the amount of care and energy required to maintain a healthy garden or landscape. Native plants are adapted to the local climate and geology, and often require less maintenance than exotic species. Native plants also support populations of native birds, insects, and other animals that they coevolved with, thus promoting a healthy community of organisms.

Plants in a garden or maintained landscape often form a source population from which plants can colonize new areas. Avoiding the use of invasive species helps to prevent such plants from establishing new populations. Similarly, the use of native species can provide a valuable source to help these plants colonise new areas. Some non-native species can form an ecological trap in which native species are lured into an environment that appears attractive but is poorly suited to them. However, in Britain research by the University of Sheffield as part of the BUGS project (Biodiversity in Urban Gardens in Sheffield) has revealed that for many invertebrates - the majority of wild animals in most gardens - it is not just native plants which can sustain them.

The findings were published in popular form in Ken Thompson's book 'No Nettles Required: The truth about wildlife gardening'. He confirms the approach which Chris Baines had promoted in 'How to Make a Wildlife Garden'.

Operational Principles

Enhancement of ecosystem services is encouraged throughout the lifecycle of any site by providing clear design, construction, (operations), and management criteria. To be sustainable over the long term requires environmental, social and economic demands are integrated to provide intergenerational equity by providing regenerative sustainable systems. Operational guidelines will link to and supplement existing guidelines for the built environment (supplementing existing green building and landscape guidelines), the wider environment, and they will include metrics (benchmarks, audits, criteria, indexes etc.) that give some measure of sustainability (a rating system) by clarifying what is sustainable or not sustainable or, more likely, what is more or less sustainable.

Scale

Impacts of a site can be assessed and measured over any spatio-temporal scale or context.

Direct and Indirect Environmental Impact

Impacts of a site may be *direct* by having direct measurable impacts on biodiversity and ecology at the site itself or *indirect* when impacts occur away from the site.

Site Principles

The following are some site principles for sustainable gardening:

- Do no harm
- Use the Precautionary principle
- Design with nature and culture
- Use a decision-making hierarchy of preservation, conservation, and regeneration
- Provide regenerative systems as intergenerational equity
- Support a living process
- Use a system thinking approach
- Use a collaborative and ethical approach
- Maintain integrity in leadership and research
- Foster environmental stewardship

Measuring Site Sustainability

One major feature distinguishing the approach of sustainable gardens, landscapes and sites from other similar enterprises is the quantification of site sustainability by establishing performance benchmarks. Because sustainability is such a broad and inclusive concept the environmental impacts of sites can be categorised in numerous ways depending on the purpose for which the figures are required. The process can include minimising negative environmental impacts and maximising positive impacts.

As currently applied the environment is usually given priority over social and economic factors which may be added in or regarded as an inevitable and integral part of the management process. A home gardener is likely to use simpler metrics than a professional landscaper or ecologist.

Three methodologies for measuring site sustainability include BREEAM developed by the BRE organisation in the UK, Leed, developed in America and the Oxford 360 degree sustainability Index used in Oxford Park and developed by the Oxford Sustainable Group in Scandinavia.

The *Sustainable Sites Initiative* is producing recommendations for the American Landscape Industry. The standards and guidelines finally adopted will lead to a uniform national standard, which does not currently exist. Sustainable Sites is currently in the pilot programme stage, and will formally introduce its first rating system by 2013.

The U.S. Green Building Council supports the project and plans to adopt the Sustainable Sites metrics into future versions of its Leadership in Energy and Environmental Design Green Building Rating System. Sites are rated according to their impact on ecosystem services: The following ecosystem services have been identified by the study group:

- Local climate regulation
- Air and water cleansing
- Water supply and regulation
- Erosion and sediment control
- Hazard mitigation
- Pollination
- Habitat functions
- Waste decomposition and treatment
- Global climate regulation
- Human health and well-being benefits
- Food and renewable non-food products
- Cultural benefits

Inputs

- Fossil fuels
- Embodied energy and water

- Compost
- Mulch
- Ecology & biodiversity
- Fertilizer
- Hard landscape materials
- Equipment
- Products

Outputs

- Energy & water
- Food
- Green waste
- Ecology & biodiversity
- Chemicals
- Old hard landscape materials
- Old equipment
- Old products

Constraints

Any kind of auditing or benchmarking will depend on the selection and weighting of the metrics chosen; the depth and detail of analysis required; the purpose for which the figures are required; and the environmental circumstances of the particular site.

Pests in Various Parts of Plants

Bulbs, Corms and Tubers

In botany, a bulb is a short stem with fleshy leaves or leaf bases that function as food storage organs during dormancy. (In gardening, other kinds of storage organ are also called "bulbs" or "ornamental bulbs".)

A bulb's leaf bases, also known as scales, generally do not support leaves, but contain food reserves to enable the plant to survive adverse conditions. At the centre of the bulb is a vegetative growing point or an unexpanded flowering shoot. The base is formed by a stem, and plant growth occurs from this basal plate. Roots emerge from the underside of the base, and new stems and leaves from the upper side. Tunicate bulbs have dry, membranous outer scales that protect the continuous lamina of fleshy scales. Species in the genera *Allium*, *Hippeastrum*, *Narcissus*, and *Tulipa* all have tunicate bulbs. Non-tunicate bulbs, such as *Lilium* and *Fritillaria* species, lack the protective tunic and have looser scales. Other types of storage organs (such as corms, rhizomes, and tubers) are sometimes referred to as bulbs, although as the term is used in botany, they are not. The technical term for plants that form underground storage organs, including bulbs as well as tubers and corms, is geophyte. Some epiphytic orchids (family Orchidaceae) form above-ground storage organs called pseudobulbs, that superficially resemble bulbs.

Nearly all plants that form true bulbs are monocotyledons, and include:

- *Amaryllis*, *Hippeastrum*, *Narcissus*, and several other members of the amaryllis family Amaryllidaceae. This includes onion, garlic, and other alliums, members of the Amaryllid subfamily Allioideae.
- Lily, tulip, and many other members of the lily family Liliaceae.
- Two groups of *Iris* species, family Iridaceae: subgenus *Xiphium* (the "Dutch" irises) and subgenus *Hermodactyloides* (the miniature "rock garden" irises).

Oxalis, in the family Oxalidaceae, is the only dicotyledon genus that produces true bulbs.

Bulbous plant species cycle through vegetative and reproductive growth stages; the bulb grows to flowering size during the vegetative stage and the plant flowers during the reproductive stage. Certain environmental conditions are needed to trigger the transition from one stage to the next, such as the shift from a cold winter to spring. Once the flowering period is over, the plant enters a foliage period of about six weeks during which time the plant absorbs nutrients from the soil and energy from the sun for setting flowers for the next year. Bulbs dug up before the foliage period is completed will not bloom the following year but then should flower normally in subsequent years.

After the foliage period is completed, bulbs may be dug up for replanting elsewhere. Any surface moisture should be dried, then the bulbs may be stored up to about 4 months for a fall planting. Storing them much longer than that may cause the bulbs to dry out inside and become nonviable.

Bulbil

A bulbil is a small bulb, and may also be called a bulblet, bulbet, or bulbel.

Small bulbs can develop that replace or propagate a large bulb. If one or several moderate-sized bulbs form to replace the original bulb, they are called renewal bulbs. Increase bulbs are small bulbs that develop either on each of the leaves inside a bulb, or else on the end of small underground stems connected to the original bulb.

Some lilies form small bulbs, called bulbils, in their leaf axils. Several members of the onion family, Alliaceae, including *Allium sativum* (garlic), form bulbils in their flower heads, sometimes as the flowers fade, or even instead of the flowers. The so-called tree onion (*Allium cepa* var. *proliferum*) forms small onions which are large enough for pickling.

Some ferns, such as Hen and Chicken Fern produce new plants at the tips of the fronds' pinnae, which are sometimes referred to a bulbils.

Bulb Pests and Diseases

As with all garden plants bulbs are susceptible to attack by certain garden pests. If left unchecked these pests can cause huge amounts of disease and damage. Resolva Bug Killer can be applied all year round for the control of aphids, whitefly, thrips and mealy bugs It also contains a foliar feed that provides vital support to your plants while the pest problem is being solved. Resolva Bug Killer kills on contact meaning troublesome pests are dealt with quick and easily. For larger areas it is also available in concentrate formulation.

Slugs can also cause damage to bulbs, particularly tulips, gladioli, dahlias and lilies. For slugs choose to use Eraza Slug and Snail Killer at the first sign of trouble and supplement with the barrier Earth Matters Slug Blocker Granules. Eraza Slug and Snail Killer offer a complete and effective control for slugs and snails. The pellets are rain proof and pet repellent so that if applied at the recommended rate they have longer staying power without posing a risk to pets and wildlife.

Roots

The root is the organ of a plant that typically lies below the surface of the soil. However, roots can also be aerial or aerating (growing up above the ground or especially above water). Furthermore, a stem normally occurring below ground is not exceptional either.Therefore, the root is best defined as the non-leaf, non-nodes bearing parts of the plant's body. However, important internal structural differences between stems and roots exist.

The first root that comes from a plant is called the radicle. The four major functions of roots are 1) absorption of water and inorganic nutrients, 2) anchoring of the plant body to the ground, and supporting it, 3) storage of food and nutrients, 4) vegetative reproduction. In response to the concentration of nutrients, roots also synthesise cytokinin, which acts as a signal as to how fast the shoots can grow. Roots often function in storage of food and nutrients. The roots of most vascular plant species enter into symbiosis with certain fungi to form mycorrhizae, and a large range of other organisms including bacteria also closely associate with roots.

Root Growth

Early root growth is one of the functions of the apical meristem located near the tip of the root. The meristem cells more or less continuously divide, producing more meristem, root cap cells (these are sacrificed to protect the meristem), and undifferentiated root cells. The latter become the primary tissues of the root, first undergoing elongation, a process that pushes the root tip forward in the growing medium. Gradually these cells differentiate and mature into specialized cells of the root tissues.

There is correlation of roots using the process of plant perception to sense their physical environment to grow, including the sensory of light, and physical barriers. Plant roots will generally grow in any direction where the correct environment of air, mineral nutrients and water exists to meet the plant's needs. Roots will shy or shrink away from dry, or other poor soil conditions.

Over time, given the right conditions, roots can crack foundations, snap water lines, and lift sidewalks. At germination, roots grow downward due to gravitropism, the growth mechanism of plants that also causes the shoot to grow upward. In some plants (such as ivy), the "root" actually clings to walls and structures.

Growth from apical meristems is known as primary growth, which encompasses all elongation. Secondary growth encompasses all growth in diametre, a major component of woody plant tissues and many nonwoody plants. For example, storage roots of sweet potato have secondary growth but are not woody. Secondary growth occurs at the lateral meristems, namely the vascular cambium and cork cambium. The former forms secondary xylem and secondary phloem, while the latter forms the periderm.

In plants with secondary growth, the vascular cambium, originating between the xylem and the phloem, forms a cylinder of tissue along the stem and root. The vascular cambium forms new cells on both the inside and outside of the cambium cylinder, with those on the inside forming secondary xylem cells, and those on the outside forming secondary phloem cells. As secondary xylem accumulates, the "girth" (lateral dimensions) of the stem and root increases. As a result, tissues beyond the secondary phloem (including the epidermis and cortex, in many cases) tend to be pushed outward and are eventually "sloughed off" (shed).

At this point, the cork cambium begins to form the periderm, consisting of protective cork cells containing suberin. In roots, the cork cambium originates in the pericycle, a component of the vascular cylinder. The vascular cambium produces new layers of secondary xylem annually. The xylem vessels are dead at maturity but are responsible for most water transport through the vascular tissue in stems and roots.

Identifying Vegetable Root Maggots and Root Maggot Control

Vegetable root maggots are the larva of a type of fly called a root maggot fly. There are several kinds with different preferred host plants. The eggs of these root eating insects are laid in the soil and hatch into larva. The larva are the little worms you see on the roots of your plant. The larva will come to the surface to pupate and then they are adults who will start the process all over again. Eggs can survive the winter in the soil. If a plant is unexplainably stunted or if it starts to wilt for no reason, there may be vegetable root maggots in the soil. Root maggots are more likely to attack in cool weather.

The best way to tell is to gently lift the plant from the soil and examine their roots. If vegetable root maggots are the culprit, the roots will be eaten away or tunneled through in the case of larger rooted plants like turnips. And, of course, root maggot larva will be present.

Root maggots typically attack either legume plants (beans and peas) or cruciferous plants (cabbage, broccoli, turnips, radishes, etc.) but they are not exclusive to those plants and can be found on almost any type of vegetable.

Root Maggot Control

These root eating insects will stay in your garden beds and attack other plants unless you take steps to get rid of them. There are several things you can do for root maggot control.

The first thing to do is to get rid of infested plants. Dying plants will attract the root maggot fly and should either be disposed of in the trash or burned. Do not compost them. Once a plant is infested, it cannot be saved, but you can do several things to keep the next plants from becoming infected.

Organic root maggot control can be:

- Dusting the plants with diatomaceous earth
- Adding beneficial nematodes to the soil
- Releasing predatory rove beetles into your garden
- Covering plants with floating row covers
- Solarizing infected beds

If you wish to use chemicals for root maggot control, apply a liquid pesticide to your garden bed at the beginning of the growing season. Make sure that you soak the soil. This will kill off the vegetable root maggots. Keep in mind that anything else in treated soil, such as worms, will also be killed.

Root Weevil

Root weevils are a plant pest both indoors and outdoors. These destructive little insects will invade the root system of a healthy plant and then proceed to eat the plant from the roots up. Identifying and controlling root weevil in your garden and houseplants can keep your plants from suffering unnecessary damage.

Root weevils can be one of several kinds. The most common in the garden is the black vine root weevil or the strawberry root weevil. The black vine weevil attacks shrubs and strawberry weevils attack strawberries. While these are the most common, they are far from the only kind. All plants in your home or garden are susceptible to weevil infestation.

Larval root weevils will look like white grubs or worms and will be found in the soil. Adult weevils are beetle like insects that can be black, brown or grey. If root weevils are present in your garden or houseplants, there will be damage to both the roots and the leaves. The leaves of the plant will be irregular, as though someone has been taking bites out of the edges. This damage will appear in the night, as root weevils come out to feed at night.

Controlling root weevil is possible. Organic root weevil control methods include purchasing parasitic nematodes or predatory beetles can be bought to hunt down the weevils. You can also hand pick the adults off the plant at night while they are eating. Weevils are also attracted to moisture, so a shallow pan of water can be set out at night and the weevils will climb into it and drown.

Non-organic root weevil control methods are to spray the leaves of the plant with a pesticide and to thoroughly soak the soil with a liquid pesticide. But, remember, when you do this, you may also be killing off beneficial insects and small animals as well.

Finding these insects in the roots and leaves of your plants is unpleasant, but it can be fixed. As always, the best root weevil control is to make sure you never get any in the first place. Be sure to practice good garden hygiene and clean up dead plants and do not over mulch.

Stems and Shoots

A stem is one of two main structural axes of a vascular plant, the other being the root. The stem is normally divided into nodes and internodes. The nodes hold buds which grow into one or more leaves, conifer cones, roots, other stems, or flowers (inflorescences); the internodes distance one node from another. The term "shoots" is often confused with "stems"; "shoots" generally refers to new fresh plant growth including both stems and other structures like leaves or flowers. In most plants stems are located above the soil surface but some plants have underground stems. A stem develops buds and shoots and usually grows above the ground. Inside the stem, materials move up and down the tissues of the transport system.Stems are often green.Stems have thin pipes to carry water around the plant.Stem consists of xylem and phloem tissue. Stems and roots hold the plant upright.

Buds

In botany, shoots consist of stems including their appendages, the leaves and lateral buds, flowering stems and flower buds. The new growth from seed germination that grows upward is a shoot where leaves will develop. In the spring, perennial plant shoots are the new growth that grows from the ground in herbaceous plants or the new stem and/or flower growth that grows on woody plants.

Figure: *Edible shoots of Sachaline*

In everyday speech, shoots are often synonymous with stems. Stems, which are an integral component of shoots, provide an axis for buds, fruits, and leaves.

Young shoots are often eaten by animals because the fibres in the new growth have not yet completed secondary cell wall development, making the young shoots softer and easier to chew and digest. As shoots grow and age, the cells develop secondary cell walls that have a hard and tough structure. Some plants (e.g. bracken) produce toxins that make their shoots inedible or less palatable.

Leaf

A leaf is an organ of a vascular plant, and is the principal appendage of the vascular plant stem. The leaves and stem together form the shoot. Foliage is a mass noun that refers to leaves collectively.

Typically a leaf is a thin, dorsiventrally flattened organ, borne above ground and specialized for photosynthesis. Most leaves have distinctive upper (adaxial) and lower (abaxial) surfaces that differ in colour, hairiness, the number of stomata and other features. The palisade mesophyll almost always occurs on the upper side of the blade or lamina of the leaf but in some species, including the mature foliage of *Eucalyptus* palisade occurs on both sides and the leaves are said to be isobilateral. Many types of leaves are adapted in ways almost unrecognisable in those terms: some are not flat (for example many succulent leaves and the needles of conifers), some are not above ground (such as bulb scales), and some are without major photosynthetic function (consider for example cataphylls, and spines). Furthermore, several kinds of leaf-like structures found in vascular plants are not totally homologous with them. Examples include phyllodes, cladodes, and phylloclades that differ from leaves in their structure and origin. Conversely, many structures of non-vascular plants, and even of some lichens, which are not plants at all (in the sense of being members of the kingdom Plantae), do look and function much like leaves.

Leafminers

Leafminers are pests which tunnel through the mesophyll of leaves creating distinctive and unsightly squiggly patterns. They are, collectively, mostly comprised of members of two insect orders: Lepidoptera (moths and butterflies) and Diptera (flies). The latter will be the focus of this page as caterpillar pests are covered in the previous page. There are about 98,000 members of the order Diptera, and many are pests in some way, shape or form, but only a handful are active as leafminers. In the United States they are not as common as pests like aphids and caterpillars, but do deserve attention. Two decades ago, and we don't know why, they were a more serious problem than they are today, but are still a concern for growers and, thus, mentioned on this website.

How They Become a Problem

Leafminers can attack outdoor and indoor crops of many kinds, but the damage they cause is not usually life-threatening. Aesthetically speaking, though, these flies can destroy the value of a crop. Larger plants, shrubs and trees can be infested by leafminers every year without noticeable growth setbacks unless the infestation is serious-in which case photosynthesis can be hindered. And cosmetic issues are not always a concern unless the plants, shrubs or tree are to be sold outright. Growers' concerns usually involve leafminer infestations on smaller specimens which will be marketed as plants for their appearance and not for their fruits, etc. Chrysanthemums are one example of such a crop. On mums and other such flowering plants, leafminers are a more serious concern.

A Serious Pest

The infestation process begins as the female flies lands on your plants and begin to feed. They do this by puncturing the leaf with their sucking mouth parts. This activity creates a tiny white spot which, by itself, is not of great concern to growers-though a lot of feeding activity can make the plant's leaves unsightly. The real damage begins when the female fly lays an egg in the puncture wound. The egg hatches and the resulting larva or maggot begins to tunnel through the leaf, eating the cells of which the mesophyll is composed, creating a void, tunnel or mine as they go. The outer leaf surfaces remain intact, but with no cells between the upper and lower surfaces the feeding becomes very noticeable and ugly. As the maggot continues to feed, it tunnels randomly. Moreover, the maggot increases in size so the mines get larger and more noticeable as progress is made. Additionally, the maggot's feces is forced out of the mine's entrance creating yet another unsightly mess (and this is exacerbated when a maggot has been parasitized). The growth process continues until the maggot is ready, through complete metamorphosis, to pupate. At this point they usually drop to the ground. On the ground, control of these pests can obtained by employing one or more soil-pest controls. When the entire process is complete, the pupae give forth a new generation of adult flies ready to mate and lay eggs in leaves so that the cycle may continue.

For the Scout

You will probably not be bothered with large numbers of leafminers and, unless they are a known problem-pest in your area, we wouldn't recommend special treatments or scouting measures be taken. Become visually familiar with them-which can be difficult as they are unremarkable in appearance and look like, well, little two to three millimetre flies with a yellow spot on each side of their thorax, in essence, stay alert and do nothing else. Utilizing yellow sticky traps during a regular scouting regimen should be the only step you'd take in an effort to scout for them, aside, that is, from looking for those telltale mines during your normal plant inspections. Which, by the way, if you do find mined leaves, remove and destroy them.

Some Solutions

There are few biological pest control agents available to specifically address leafminer pests from the order Diptera. The parasitoids which are available have been known to do a very good job, though. If possible, use them preventively or, at least, at the very first sign of these flies. Other biocontrol agents, those used for soil pests, for example, can also help you with leafminers while being used primarily to thwart other pests like fungus gnats. Chemical controls can also be effective, but expect difficulty when trying to effectively treat the leaf-protected larvae and soil-protected pupae.

Flowers

A flower, sometimes known as a bloom or blossom, is the reproductive structure found in flowering plants (plants of the division Magnoliophyta, also called angiosperms). The biological function of a flower is to effect reproduction, usually by providing a mechanism for the union of sperm with eggs. Flowers may facilitate outcrossing (fusion of sperm and eggs from different individuals in a population) or allow selfing (fusion of sperm and egg from the same flower). Some flowers produce diaspores without fertilization.

Common Pests: The most common pests of foliage and flowering plants are spider mites, mealy bugs, fungus gnats, whitefly and aphids. The use of plant material for outdoor as well as indoor decoration exposes them to all manner of garden pests, such as caterpillars, slugs, snails and thrips, many of which thrive once the plants are moved back indoors. For this reason, plants which are being moved from decks, patios, etc. to the home or office should be washed thoroughly with a spray of clear water.

Pest Control: Despite the best control programmes employed by growers and retailers, it sometimes happens that a new plant harbors pests whose eggs have survived to hatch in you home conditions. A period of isolation (10-14 days) may be the answer. Yet some pests may not show up for a longer time, being present in small numbers until the indoor climate becomes more favourable; for example, spider mites thrive in dry air, so they may not become evident until the humid season is past.

Regular and frequent checking for abnormal growth and insects can be combined with your checks for watering needs. Most pests lurk under leaves and close to the growing tip where tissue is softest. Most of these pests can be washed off with a spray of water. Several washes over a period of two or three weeks, plus isolation for the infested plant to prevent the pest from moving to other plants, generally eliminates the problem. Chemical sprays are not pleasant to use, and are generally recommended for outdoor application.. Some pesticides are hazardous, not only to the general environment, but to the plant tissue. Ferns are extremely sensitive to all forms of chemical spray, and infested fronds are best removed and destroyed.

Diseases: Diseases are rare on foliage and flowering plants. The only ones likely to be encountered are Botrytis, the gray mold fungus that invades dead and dying tissues, and soft leaf spot or crown rot fungi which thrive in moist atmospheres. The best cure is to remove affected parts, and avoid splashing the plant when watering; give more space for air circulation, or move infected plants to a drier place.

Fruit

In botany, a fruit is a part of a flowering plant that derives from specific tissues of the flower, one or more ovaries, and in some cases accessory tissues. Fruits are the means by which these plants disseminate seeds. Many of them that bear edible fruits, in particular, have propagated with the movements of humans and animals in a symbiotic relationship as a means for seed dispersal and nutrition, respectively; in fact, humans and many animals have become dependent on fruits as a source of food.

Fruits account for a substantial fraction of the world's agricultural output, and some (such as the apple and the pomegranate) have acquired extensive cultural and symbolic meanings.

In common language usage, "fruit" normally means the fleshy seed-associated structures of a plant that are sweet or sour and edible in the raw state, such as apples, oranges, grapes, strawberries, bananas, and lemons. On the other hand, the botanical sense of "fruit" includes many structures that are not commonly called "fruits", such as bean pods, corn kernels, wheat grains, and tomatoes.

The section of a fungus that produces spores is also called a fruiting body.

Natural predators such as lacewings, leatherwing beetles, ladybird beetles, ground beetles, wasps, praying mantis and pirate bugs will control or contain most pest populations at an acceptable level, especially if trees are kept vigorous, orchard areas are kept clean of trash and weeds, and trees are well pruned to facilitate good air movement. Home orchardists have little need to completely eradicate pests.

Most insect and mite pests of fruit trees are controlled by many beneficial species of insects and mites found in the orchard. Do not spray pests unless you are certain they are present in damaging numbers or this publication suggests you do so. Unnecessary sprays reduce control provided by beneficial species and may result in added damage from pests freed from their natural controls.

Several of the most common fruit tree diseases may be controlled by using the proper fixed copper spray during the dormant season. These diseases include: bacterial canker, brown rot, coryneum blight and peach leaf curl.

How Pests, Diseases, and Hosts Fit Together

The tables, below, are brief formats designed only as guides to be used with Integrated Pest Management (IPM) and other research-based references. If one or more of these pest and disease problems are not present, do not apply preventatives. Many do not become problems until trees begin to bear.

How Pests, Diseases, and Hosts Fit Together

Pest/Disease	*Hosts(s)*
Codling Moth	Apple, pear, sometimes walnut, plum
Peach Twig Borer (PTB)	Peach & nectarine, almond, apricot, plum & prune
Brown Rot Fungus	Peach & nectarine, cherry, plum & prune, almond
Fireblight	Pear, apple, quince
Scab	Apple, pear
Bacterial Canker	Peach & nectarine, cherry, plum & prune, pear (blast)
Peach Leaf Curl	Peach & nectarine

Goals for reducing insect and mite damage in home orchards are to:

- Protect trees and crops from substantial damage
- Make it simple and easy
- Make it safe
- Use methods that enhance natural control factors

Dormant season treatments are most important since they will not kill beneficial insects. One spray controls many orchard pests. The following tables of Orchard Pests by Season list several common home orchard pests and include information on damage, description, life cycle, and control options. More importantly, it provides information on the seasons when these pests should be addressed if they are problems in the home orchard. When damage becomes obvious, it is often too late to treat with successful results for that year.

Seeds

A seed is an embryonic plant enclosed in a protective outer covering called the seed coat, usually with some stored food. It is a characteristic of spermatophytes (gymnosperm and angiosperm plants) and the product of the ripened ovule which occurs after fertilization and some growth within the mother plant. The formation of the seed completes the process of reproduction in seed plants (started with the development of flowers and pollination), with the embryo developed from the zygote and the seed coat from the integuments of the ovule.

Seeds have been an important development in the reproduction and spread of gymnosperm and angiosperm plants, relative to more primitive plants such as ferns, mosses and liverworts, which do not have seeds and use other means to propagate themselves. This can be seen by the success of seed plants (both gymnosperms and angiosperms) in dominating biological niches on land, from forests to grasslands both in hot and cold climates.

The term "seed" also has a general meaning that antedates the above—anything that can be sown, e.g. "seed" potatoes, "seeds" of corn or sunflower "seeds". In the case of sunflower and corn "seeds", what is sown is the seed enclosed in a shell or husk, whereas the potato is a tuber.

Many structures commonly referred to as "seeds" are actually dry fruits. Plants producing berries are called baccate. Sunflower seeds are sometimes sold commercially while still enclosed within the hard wall of the fruit, which must be split open to reach the seed. Different groups of plants have other modifications, the so-called stone fruits (such as the peach) have a hardened fruit layer (the endocarp) fused to and surrounding the actual seed. Nuts are the one-seeded, hard-shelled fruit of some plants with an indehiscent seed, such as an acorn or hazelnut.

Seed Treatment

In agriculture and horticulture, a seed treatment or seed dressing is a chemical, typically antimicrobial or fungidal, with which seeds are treated (or "dressed") prior to planting. Less frequently insecticides are added. Seed treatments can be an environmentally more friendly way of using pesticides as the amounts used can be very small. It is usual to add colour to make treated seed less attractive to birds if spilt and easier to see and clean up in the case of an accidental spillage.

One seed treatment, imidacloprid, from the neonicotinoid family of insecticides, is controversial and was banned in France for use on maize, due to that government's belief that the chemical was implicated in recent dramatic drops in bee counts, and possibly in Colony Collapse Disorder. Dust from treated seed is known to have caused at least some problems particularly from crops such as maize drilled during the main honey flows. Improvements to pneumatic drills to reduce dust release, and improvements to seed treatment compounds to prevent the compound breaking up into dust have been introduced in

Europe led by Germany and Holland from 2009 to 2012. Information on seed treatments including the information above can be seen on the registration authority databases.

Seed coating is a thicker form of covering of seed and may contain fertiliser, growth promoters and or seed treatment as well as an inert carrier and a polymer outer shell. Seed dressing is also used to refer to the process of removing chaff, weed seeds and straw from a seed stock. Care is needed not to confuse the two.

Insect Pests of Malvaceous Crops

The principle economic use of Malvaceous plants is as a source of natural fibres, the family providing perhaps the world's 3 most important fibre crops. Plants of the family are also used for food, beverages, timber, in traditional medicine, and in horticulture.

Fibre

The most important crop is cotton, which is obtained from the seed hairs of 4 species of *Gossypium*, the Old World diploids *Gossypium arboreum* and *Gossypium herbaceum*, and the New World tetraploids *Gossypium hirsutum* and *Gossypium barbadense*. Cotton is the world's preeminent natural fibre, of which in excess of 20 million tonnes are produced annually. This dwarfs the production of animal fibres (wool, about 1½ million tonnes annually) and is comparable to the total production of synthetic fibres.

The largest producers are China, the United States of America, India and Pakistan. Production is also extensive in Central Asia, the Middle East, Australia, Brasil, and Mediterranean Europe.

Jute, obtained from the phloem fibres of *Corchorus capsularis* and *Corchorus olitorius*, is also an important fibre crop; with a production in excess of 2 million tonnes annually it is the second largest vegetable fibre. India and Bangladesh are by far the largest producers. Kenaf, obtained from the phloem fibres of *Hibiscus cannabinus*, has an annual production of about 1 million tonnes. Thailand, Burma, China and Bangladesh are the major producers.

Several other species are exploited for their phloem fibres, including devil's cotton, *Abroma angusta*, China jute, *Abutilon theophrastii*, roselle, *Hibiscus sabdariffa* and bast, *Tilia* sp., and many others are or were used at the subsistence level.

Kapok, the seed hairs of a number of trees belonging to Bombacoideae, particularly *Bombax ceiba* (250,000 tonnes annually), is not spinnable and therefore can't be used for textiles, but is waterproof. It is used primarily for insulation.

Food

The most important food crops are durian, the fruit of *Durio zibethinus*, which is an important plantation crop in South East Asia, okra, the fruit of *Abelmoschus esculentus*, which is used as a vegetable, and cacao, the fruit of *Theobroma cacao*, which is the source of chocolate and cacao butter. The fruits of a number of other species are also eaten. Several species, including *Abelmoschus manihot* and species of *Corchorus* and *Malva*, are grown as leaf vegetables. Marshmallows were originally made from the candied roots of *Althaea officinalis*.

Vegetable Oils

Vegetable oil is produced as a byproduct (cottonseed oil, kenaf oil) of the cultivation of cottons and kenaf for fibre. A vegetable oil is also produced from the baobab (*Adansonia*).

Beverages

Kola nuts, the fruits of *Cola nitida* and other species is used commercially as a flavouring ingredient in drinks (colas or kolas). They contain stimulant alkaloids - caffeine and theobromine. Cacao is used in the production of beverages, in addition to chocolate.

Timber

Many arborescent species of Malvaceae are used for timber, some of them commercially. Perhaps the most important are *Triplochiton*, *Mansonia* and *Heritiera*, but *Tilia*, *Talipariti* and many other genera are used. Balsa, *Ochroma pyramidale*, is noted for the low density of its wood.

Medicinal

Many species of Malvoideae have been exploited for the demulcent nature of their mucilaginous sap. The best known source is marsh-mallow, *Althaea officinalis*.

Horticulture

Many species of Malvaceae have showy flowers, and have a place in horticulture, the number of species used in at least a small way exceeding one hundred. In temperate regions the predominant malvaceous horticultural subjects are hollyhocks, *Alcea rosea*, musk mallows, *Malva moschata*, shrub Lavateras, *Lavatera ×clementii*, annual Lavateras, *Lavatera trimestris*, hardy hibiscus, *Hibiscus moscheutos* and allies, and Rose of Sharon, *Hibiscus syriacus*.

Biological Control

Biological control represents one alternative to the use of insecticides. Biological control is the conscious use of living beneficial organisms, called natural enemies, for the control of pests. Virtually all pests have natural enemies and appropriate management of natural enemies can effectively control many pests. The natural enemies of insects are a diverse group of organisms that includes predators, parasitic insects, nematodes, and various mircoorganisms. Although biological control will not control all pests all of the time, it is a key component of integrated pest management. The intent of biological control is not to eradicate pests, but to keep them at tolerable levels at which they cause no appreciable harm.

In the upper Midwest, literally hundreds of types of natural enemies occur naturally in the environments of agricultural crops, managed forests, and managed urban and suburban landscapes. These have a major impact on pest populations, and often result in pests being kept below damaging levels. When there is no human activity involved inmanaging these natural enemies, we refer to the results as "natural control" of the pests. Natural control is of major importance in pest management, but the benefits frequently go unrecognized because the natural enemies are often very small and their activity is easily overlooked.

To get the greatest benefit from both natural control and biological control, it is necessary to recognize natural enemies, understand their biological characteristics (such as life cycle, preferred host insects, and efficiency at controlling pests). Some natural enemies, such as lady beetles, are readily recognized in the adult stage by most people but may not be recognized in the egg, larval, or pupal stages. Other natural enemies, including most parasites, are much more difficult to recognize.

Although birds, mammals, frogs, and other higher animals can be important as natural enemies, they can rarely be effectively managed for biological control. These animals lie outside the scope of this discussion, which deals primarily with the predatory or parasitic insects of pest insects and mites.

Insect Pests of Cole Crops

Cole crops are a common sight in the home garden, especially in cooler weather. But some gardeners may not know what cole crops are. Whether you know what cole crop plants are or not, chances are you enjoy them on a regular basis.

Cole crops, at a basic level, are plants that belong to the mustard (Brassica) family and are all descendants of wild cabbage. As a group, these plants grow better in cool weather. This leads many people to think that the word "cole" is a variation of the word "cold" and they may even refer to these plants as cold crops. Actually, the word "cole" is a variation of a Latin word that means stem.

The following is a list of the most common of these plants:

- Brussels sprout
- Cabbage
- Cauliflower
- Collards
- Kale
- Kohlrabi
- Mustard
- Broccoli
- Turnip
- Watercress

When to Plant Cole Crops

The specific time when to plant cole crops will be different depending on which one you are growing. For example, most cabbage varieties can be planted out much earlier than broccoli or cauliflower because cabbage plants can tolerate a much lower temperature. In general, these crops grow best when daytime temperatures are below 80 degrees F. and nighttime temperatures are below 60 degrees F. at night. Temperatures higher than this can lead to buttoning, bolting or poor head formation.

Growing Cole Crop Plants

For best results, cole crops should be grown in full sun, but due to their need for cooler temperatures, if you have a partially shaded garden, the vegetables in this family will do okay here as well. Also, if you live in an area that has a short cool season, planting them in part shade can help alleviate daytime temperatures by keeping direct sun from falling on the plants.

Cole crop plants usually need significant amounts of nutrients, particularly micro-nutrients that may not be found in standard fertilizers. Therefore, it is important to work organic material into the beds you plan on growing cole crops in prior to planting them.

Since many of these crops are susceptible to the same types of diseases and pests, rotating plants at least every few years is a good idea. This will help to cut down on diseases and pests that overwinter in the soil and attack the plants.

Insect Pests of Solanaceous Vegetables

The Solanaceae, or nightshades, is an economically important family of flowering plants. The family ranges from annual and perennial herbs to vines, lianas, epiphytes, shrubs and trees, and includes a number of important agricultural crops, medicinal plants, spices, weeds, and ornamentals. Many members of the family contain potent alkaloids, and some are highly toxic, but many cultures eat nightshades, in some cases as staple foods. The family belongs to the order Solanales, in the asterid group dicotyledons (Magnoliopsida). The Solanaceae consists of approximately 98 genera and some 2,700 species, with a great diversity of habitats, morphology and ecology.

The name Solanaceae derives from the genus *Solanum*, "the nightshade plant". The etymology of the Latin word is unclear. The name may come from a perceived resemblance of certain solanaceous flowers to the sun and its rays. At least one species of *Solanum* is known as the "sunberry". Alternatively, the name could originate from the Latin verb *solari*, meaning "to soothe", presumably referring to the soothing pharmacological properties of some of the psychoactive species of the family.

The family has a worldwide distribution, being present on all continents except Antarctica. The greatest diversity in species is found in South America and Central America.

The Solanaceae include a number of commonly collected or cultivated species. The most economically important genus of the family is *Solanum*, which contains the potato (*Solanum tuberosum*, in fact, another common name of the family is the "potato family"), the tomato (*Solanum lycopersicum*), and the eggplant or aubergine (*Solanum melongena*). Another important genus, *Capsicum*, produces both chili peppers and bell peppers.

The genus *Physalis* produces the so-called groundcherries, as well as the tomatillo (*Physalis philadelphica*), the Cape gooseberry and the Chinese lantern. The genus *Lycium* contains the boxthorns and the wolfberry *Lycium barbarum*. *Nicotiana* contains, among other species, the plant that produces tobacco. Some other important members of Solanaceae include a number of ornamental plants such as *Petunia*, *Browallia* and *Lycianthes*, the source of psychoactive alkaloids, *Datura*,

Mandragora (mandrake), and *Atropa belladonna* (deadly nightshade). Certain species are universally known for their medicinal uses, their psychotropic effects, or for being poisonous.

With the exception of tobacco (*Nicotiana tabacum*, Nicotianoideae) and petunia (*Petunia x hybrida*, Petunioideae), most of the economically important genera are contained in the subfamily Solanoideae. Finally, but not less importantly, the Solanaceae include many model organisms which are important in the investigation of fundamental biological questions at cellular, molecular, and genetic levels, such as tobacco and the petunia.

Description

Plants in the Solanaceae can take the form of herbs, shrubs, trees, vines and lianas and sometimes epiphytes. They can be annuals, biennials or perennials, upright or decumbent. Some have subterranean tubers. They do not have laticifers, nor latex, nor coloured saps. They can have a basal or terminal group of leaves or neither of these types. The leaves are generally alternate or alternate to opposed (that is, alternate at the base of the plant and opposed towards the inflorescence). The leaves can be herbaceous, leathery or transformed into spines.

The leaves are generally petiolate or subsessile, rarely sessile. They are frequently inodorous, but on occasions they are aromatic or fetid. The foliar lamina can be either simple or compound, the latter can be either pinnatifid or ternate. The leaves have reticulated venation and lack a basal meristem. The laminae are generally dorsiventral and lack secretory cavities. The stomata are generally confined to one of a leaf's two sides, they are rarely found on both sides.

The flowers are generally hermaphrodite, although some are monoecious, andromonoecious, or dioecious species (such as some *Solanum* or *Symonanthus*). Pollination is entomophilous. The flowers can be solitary or grouped into terminal, cymose, or axillary inflorescences. The flowers are medium-sized, fragrant (*Nicotiana*), fetid (*Anthocercis*) or inodorous. The flowers are usually actinomorphic, slightly zygomorphic, or markedly zygomorphic (for example, in flowers with a bilabial corolla in *Schizanthus* species).

The irregularities in symmetry can be due to the androecium, to the perianth, or both at the same time. In the great majority of species, the flowers have a differentiated perianth with a calyx and corolla (with five sepals and five petals, respectively) an androecium with five stamens and two carpels forming a gynoecium with a superior ovary (they are therefore referred to as pentamers and tetracyclic). The stamens are epipetalous and are typically present in multiples of four or five, most commonly four or eight. They usually have a hypogynous disk. The calyx is gamosepalous (as the sepals are joined together forming a tube), with the (4)5(6) segments equal, it has five lobes, with the lobes shorter than the tube, it is persistent and often accrescent.

The corolla usually has five petals that are also joined together forming a tube. Flower shapes are typically rotate (wheel-shaped, spreading in one plane, with a short tube) or tubular (elongated cylindrical tube), campanulated or funnel-shaped.

The androecium has (2)(4)5(6) free stamens within it, oppositsepals (that is, they alternate with the petals), they are usually fertile or, in some cases (for example in Salpiglossideae) they have staminodes. In the latter case, there is usually either one staminode (*Salpiglossis*) or three (*Schizanthus*).

The anthers touch on their upper end forming a ring, or they are completely free, dorsifixed or basifixed with poricide dehiscence or through small longitudinal cracks. The stamen's filament can be filliform or flat. The stamens can be inserted inside the coralline tube or exserted. The plants demonstrate simultaneous microsporogenesis, the microspores are tetrad, tetrahedral or isobilateral. The pollen grains are bicellular at the moment of dehiscence, usually open and angular.

The gynoecium is bicarpelar (rarely three- or five-locular) with a superior ovary and two locules, which may be secondarily divided by false septa, as is the case for Nicandreae and Datureae. The gynoecium is located in an oblique position relative to the flower's median plane. They have one style and one stigma; the latter is simple or bilobate. Each locule has one to 50 ovules that are anatropous or hemianatropous with axillar placentation.

The development of the embryo sack can be the same as for *Polygonum* or *Allium* species. The embryo sack's nuclear poles become fused before fertilization. The three antipodes are usually ephemeral or persistent as in the case of *Atropa*.

The fruit can be a berry as in the case of the tomato or wolfberry, a dehiscent capsule as in *Datura*, or a drupe. The fruit has axial placentation. The capsules are normally septicidal or rarely loculicidal or valvate.

The seeds are usually endospermic, oily (rarely starchy), and without obvious hairs. The seeds of most Solanaceae are round and flat, about 2–4 mm (0.079–0.157 in) in diametre. The embryo can be straight or curved and has two cotyledons. Most species in the Solanaceae have 2n=24 chromosomes, but the number may be a higher multiple of 12 due to polyploidy.

Wild potatoes, of which there are about 200, are predominantly diploid ($2 \times 12 = 24$ chromosomes), but triploid ($3 \times 12 = 36$ chromosomes), tetraploid ($4 \times 12 = 48$ chromosomes), pentaploid ($5 \times 12 = 60$) and even hexaploid ($6 \times 12 = 72$ chromosome) species or populations exist. The cultivated species *Solanum tuberosum* has $4 \times 12 = 48$ chromosomes. Some *Capsicum* species have $2 \times 12 = 24$ chromosomes, while others have 26 chromosomes.

The Diversity of Some Characteristics

Despite the previous description, the Solanaceae exhibit a large morphological variability, even in their reproductive characteristics. Examples of this diversity include:

The number of carpels that form the gynoecium: In general, the Solanaceae have a gynoecium (the female part of the flower) formed of two carpels. However, *Melananthus* has a monocarpelar gynoecium, there are three or four carpels in *Capsicum*, three to five in *Nicandra*, some species of *Jaborosa* and *Trianaea* and four carpels in *Iochroma umbellatum*.

The number of locules in the ovary: The number of locules in the ovary is usually the same as the number of carpels. However, some species occur in which the numbers are not the same due to the existence of false septa (internal walls that subdivide each locule), such as in *Datura* and some members of the Lycieae (the genera *Grabowskia* and *Vassobia*).

Type of ovules and their number: The ovules are generally inverted, folded sharply backwards (anatropous), but some genera have ovules that are rotated at right angles to their stalk (campilotropous) as in *Phrodus*, *Grabowskia* or *Vassobia*), or are partially inverted (hemitropous as in *Cestrum*, *Capsicum*, *Schizanthus* and *Lycium*). The number of ovules per locule also varies from a few (two pairs in each locule in *Grabowskia*, one pair in each locule in *Lycium*) and very occasionally only one ovule is in each locule as for example in *Melananthus*.

The type of fruit: The fruits of the great majority of the Solanaceae are berries or capsules (including pyxidia) and less often drupes. Berries are common in the subfamilies Cestroideae, Solanoideae (with the exception of *Datura*, *Oryctus*, *Grabowskia* and the tribe Hyoscyameae) and the tribe Juanulloideae (with the exception of *Markea*). Capsules are characteristic of the subfamilies Cestroideae (with the exception of *Cestrum*) and Schizanthoideae, the tribes Salpiglossoideae and Anthocercidoideae, and the genus *Datura*. The tribe Hyoscyameae has pyxidia. Drupes are typical of the Lycieae tribe and in Iochrominae.

Economic Importance

The solanaceas include such important food species as the potato (*Solanum tuberosum*), the tomato (*Solanum lycopersicum*), the pepper (*Capsicum annuum*) and the aubergine or egg plant (*Solanum melongena*). *Nicotiana tabacum,* originally from South America, is now cultivated throughout the world to produce tobacco. Many solanaceas are important weeds in various parts of the world. Their importance lies in the fact that they can host pathogens or diseases of the cultivated plants, therefore their presence increases the loss of yield or the quality of the harvested product.

An example of this can be seen with *Acnistus arborescens* and *Browalia americana* that host thrips, which cause damage to associated cultivated plants, and certain species of *Datura* that play host to various types of virus that are later transmitted to cultivated solanaceas. Some species of weeds such as, for example *Solanum mauritianum* in South Africa represent such serious ecological and economic problems that studies are being carried out with the objective of developing a biological control through the use of insects.

Various solanaceas species are grown as ornamental trees or shrubs. Examples include *Brugmansia x candida* ("Angel's Trumpet") grown for its large pendulous trumpet-shaped flowers, or *Brunfelsia latifolia,* whose flowers are very fragrant and change colour from violet to white over a period of 3 days. Other shrub species that are grown for their attractive flowers are *Lycianthes rantonnetii* (Blue Potato Bush or Paraguay Nightshade) with violet-blue flowers and *Nicotiana glauca* ("Tree Tobacco") Other solanacea species and genera that are grown as ornamentals are the petunia *(Petunia* × *hybrida), Lycium, Solanum, Cestrum, Calibrachoa* × *hybrida* and *Solandra*. There is even a hybrid between *Petunia* and *Calibrachoa* (which constitutes a new nothogenus called × *Petchoa* G. Boker & J. Shaw) that is being sold as an ornamental. Many other species, in particular those that produce alkaloids, are used in pharmacology and medicine *(Nicotiana, Hyoscyamus*, and *Datura*).

Solanaceas and the Genome

Many of the species belonging to this family, among them tobacco and the tomato, are model organisms that are used for research into fundamental biological questions. One of the aspects of the solanaceas' genomics is an international project that is trying to understand how the same collection of genes and proteins can give rise to a group of organisms that are so morphologically and ecologically different. The first objective of this project was to sequence the genome of the tomato.

In order to achieve this each of the 12 chromosomes of the tomato's haploid genome was assigned to different sequencing centres in different countries. So chromosomes 1 and 10 were sequenced in the United States, 3 and 11 in China, 2 in Korea, 4 in Britain, 5 in India, 7 in France, 8 in Japan, 9 in Spain and 12 in Italy. The sequencing of the mitochondrial genome was carried out in Argentina and the chloroplast genome was sequenced in the European Union.

Key Pests of Solanaceous Vegetables

Crop	*Insect pests*	*Diseases*
Tomato	Thrips; Aphids; Mites; Whitefly; Beetles; Fruit fly; Cluster caterpillars; Looper caterpillars; Potato moth; Heliothis (Helicoverpa).	Tomato spotted wilt virus (TSWV); Tobacco mosaic virus (TMV); Bacterial canker; Bacterial speck; Bacterial spot; Damping-off; Powdery mildew; Tomato yellow leaf curl virus; Tomato leaf curl virus; Early blight; Bacterial wilt; Grey mould; Sclerotinia rot; Anthracnose; Fusarium wilt; Nematodes.
Eggplant	Aphids; Thrips; Leafhoppers; Two-spotted mite; Beetles; Whitefly; Heliothis (Helicoverpa); Ceropid (spittle) bug; Leafminers; Potato moth.	Damping-off; Tomato spotted wilt virus; Root rots; Tobacco mosaic virus; Tobamoviruses; Bacterial wilt; Nematodes.
Capsicum and chilli	Aphids; Thrips; Beetles; Queensland fruit fly; Whitefly; Mites.	Bacterial spot; Bacterial wilt; Anthracnose; Cercospora spot; Powdery mildew; Stem rot; Sudden wilt; Tomato spotted wilt virus (TSWV); Tomato yellow leaf curl virus; Capsicum chlorosis virus (CaCV); Cucumber mosaic virus (CMV); Pepper mild mottle virus (PMMV); Potato mosaic virus (OVY); Tobacco mosaic virus (TMV); Nematodes.
Potato	Thrips (Melon thrips, Western Flower thrips, Onion thrips); Aphids; Potato moth; Whitefly; Beetles; Looper caterpillars; Leafhoppers; Bugs; Potato moth.	Armillaria root rot; Bacterial wilt; Black leg; Soft rot; Potato cyst nematode; Root-knot nematode; Late blight; Tomato spotted wilt virus; Potato leaf roll virus; Potato virus Y; Common scab of potatoes; Rhizoctonia; Powdery mildew; Fusarium wilt; Anthracnose.

Pest Management

- Integrated Pest Management (IPM) – IPM methods discourage the use of broad-spectrum chemical pesticides and instead recommend growers employ a combination of plant protection strategies, such as best-farm hygiene, biological controls, and a targeted application of selective chemicals.

- Key control method considerations for viral diseases include – the use of healthy seedlings; destruction of harvested capsicum and tomato crops before planting new blocks nearby; maintenance of high crop and farm hygiene standards; and use of virus resistant varieties in disease management strategies, where possible. For crops such as capsicum that can be infected with Tobacco mosaic virus (TMV), it is important than anyone who has been smoking wash their hands before coming into contact with the crop, particularly before picking.
- Chemical control – Pests can be controlled with an appropriately registered chemical. The APVMA maintains a database of all chemicals registered for the control of pests in Australia. However, growers need to ensure that they meet the relevant Maximum Residue Limits (MRLs) for the chemicals in the end market, be it domestic or export.
- Weed management – Effective weed management strategies are limited for capsicum and chilli producers. Weed management practices include the use of plastic mulch, selective grass herbicides, hand weeding, or tillage. It is important to control weeds that may be host to a virus or insect pests.
- Control of insect pests, mites and diseases can be achieved using biological, cultural and/or chemical methods. Cost-effective, safe strategies such as irrigation at strategic times and maintaining good crop hygiene should be used. It is important to examine the history of a paddock and determine what insect pests, mites and diseases were common in previous years. Preventative measures should be taken for the pests of most concern.

Insect Pests of Mango

Hopper (*Idioscopus clypealis, Idioscopus nitidulus* and *Amritodus atkinson*) : Of all the mango pests, hoppers are considered as the most serious and widespread pest throughout the country. Large number of nymphs and adult insects puncture and suck the sap of tender parts, thereby reducing the vigour of the plants. Heavy puncturing and continuous draining of the sap causes curling and drying of the infested tissue. They also damage the crop by secreting a sweet sticky substance which facilitate the development of the fungus *Maliola managiferae,* commonly known as sooty mould affecting adversely the photosynthetic activities of the leaves. Hopper population shoots up in February-April and June-August.

Control: Three sprays of 0.15% Carbaryl or 0.04% Monocrotophos or 0.05% Phosphamidon or 0.05% Methyl Parathion are effective in controlling the hoppers. First spray should be given at the early stage of panicle formation. The second spray at full-length stage of panicles but before full bloom and the third spray after the fruits set at pea size stage are recommended. Biological control agents such as the predators *Mallada boninensis* and *Chrysopa lacciperda,* the egg parasite *Polynema* sp. and a preparation of the fungus *Beauveria bassiana* are important and useful in controlling this pest.

Mealy Bugs (Drosicha mangiferae): Nymphs and adults suck the plant sap and reduce the vigour of the plant. Excessive and continuous draining of plant sap causes wilting and finally drying of infested tissue. They also secrete honey dew, a sticky substance, which facilitate the development of the fungus *Maliola mangiferae* (sooty mould). The female adult crawls down the tree in the month of April-May and enter in the cracks in the soil for laying eggs. Just after hatching, the minute newly hatched pink to brown coloured nymphs crawl up the tree. After climbing up the tree, they start sucking the sap of tender plant parts. They are considered more important because they infest the crop during the flowering season.

Control: Flooding of orchards with water in the month of October kills the eggs. Ploughing the orchards in the month of November exposes the eggs to suns heat. Polythene bands of 400 gauge and 25 cm width fastened around the tree trunk have been found effective barrier to stop the ascent of nymphs to the trees. The band should be fastened well in advance before the hatching of eggs, i.e., during November-December. Application of 250 g per tree of Methyl Parathion 2% dust in soil around the trunk kills the newly hatched nymphs.Spraying of 0.05% Monocrotophos or 0.2% Carbaryl or 0.05% Methyl Parathion is useful in controlling young nymphs of the mealy bug. The entomogenous fungus Beauveria bassiana is found to be an effective bioagent in controlling the nymphs of this pest.

Inflorescence Midge (Erosomyia indica): The midge infests and damages the crop in three different stages. The first attack is at the floral bud burst stage. The eggs are laid on newly emerging inflorescence; the larvae tunnel the axis and thus destroy the inflorescence completely. The mature larvae make small exit holes in the axis of the inflorescence and slip down into the soil for pupation. When the tender fruits are attacked they slowly turn yellow and finally drop. The third attack is on tender new leaves encircling the inflorescence. The most damaging one is the first attack in which the entire inflorescence is destroyed even before flowering and fruiting. The inflorescence shows stunted growth and its axis bends at the entrance point of the larvae.

Control: As the larvae pupate in the soil, ploughing of the orchard expose pupating larvae to sun helps in killing them. Soil application of Heptachlor or Methyl Parathion @ 25 to 30 kg/ha kills pupating larvae in the soil. Spraying of 0.05% Fenetrothion or 0.045% Dimethoate at the bud burst stage of the inflorescence is effective.

Fruitfly (Daccus dorsalis;, D. zonatus and D. correctus): The oriental fruitfly is one of the most serious pests of mango in the country, which has created problem in the export of fresh fruits. The female punctures the outer wall of the mature fruits with the help of its pointed ovipositor and insert eggs in small clusters inside the mesocarp of mature fruits. After hatching, the larva feeds on the pulp of fruit which appears normal from outside, but drops down finally. The mature maggots fall down into the soil for pupation. The emergence of fruitfly starts from April onwards and the maximum population is recorded during May-July, which coincides with fruit maturity. The population declines slowly from August to September after that it is non-existent up to March.

Control: Bait sprays of Carbaryl (0.2%)+protein hydrolysate (0.1%) or molasses starting at first week of April and repeated once after 21 days or hanging traps containing 100 ml water emulsion of Methyl Euginol (0.1%) + Malathion (0.1%) during fruiting is effective. As per the experiments conducted in IIHR, Bangalore against fruit fly, hot water treatment at 480C for one hour gave total control in Alphonso and Totapuri.

Scale Insects: (*Pulvinanapolygonata, Aspidiotus destructor, Ceroplastis* sp. And *Rastococcus* sp): The nymphs and adult scales suck the sap of the leaves and other tender parts and reduce the vigour of the plants. They also secrete honeydew, which helps in the development of sooty mould on leaves and other tender parts of the tree. In case of severe scale infestation, growth and fruit bearing capacity of the tree is affected adversely.

Control: Pruning of the heavily infested plant parts and their immediate destruction followed by two sprays of Monocrotophos (0.04%) or Dimethoate (0.06%) at an interval of 20 days is very effective in controlling the scale population.

Shoot Borer (Chlumetia transversa): This pest is found all over the country. Larvae of this moth bore into the young shoot resulting in dropping of leaves and wilting of shoots. Larvae also bore into the inflorescence stalk. Female moths lay egg on tender leaves. After hatching, young larvae enter the midrib of leaves and then enter into young shoots through the growing points by tunneling downwards.

Control: The attacked shoots may be clipped off and destroyed. Spraying of Carbaryl (0.2%) or Quinalphos (0.5%) or Monocrotophos (0.04%) at fortnightly intervals from the commencement of new flush gives effective control of the pest. A total of 2 to 3 sprays may be given depending on the intensity of infestation.

Bark Eating Caterpillar (Indarbella quadrinotata): The old, shady and neglected orchards are more prone to attack by this pest. Larvae of this moth feed on the bark and weaken the tree. A single female lays about 300-400 eggs in batches on the bark. The caterpillar spins brown silken web on the tree, which consists of their excreta and wood particles. Larvae also make shelter tunnels inside the stem in which they rest. Larvae generally feed from April to December.

Control: Removal of the webs from tree trunks and injecting emulsion of Monocrotophos (0.05%) or DDVP (0.05%) in each hole and plugging them with mud can control the pest.

Stem Borer *(Batocera rufomaculata):* The grub of this beetle causes damage as it feeds inside the stems, boring upward resulting in drying of branches and in severe cases attacked stem is killed. Eggs are laid either in the slits of tree trunk or in the cavities in main branches and stems covered with a viscous fluid. Pupation takes place within the stem. Beetle emerges in July-August

Control: Cleaning the tunnel with a hard wire, pouring kerosene oil, creosote, petrol, crude oil or formalin and subsequently closing entrance of the tunnel with mud or plugging it with cotton wool soaked in any of the above substances kills the grub.

Shoot Gall Psylla (Apsylla cistellata): It is a very serious pest of mango in many parts of India, particularly in Tarai region of U.P., north Bihar and West Bengal. This pest creates green conical galls in leaf axis. The activity of the pest starts from August. The galls dry out after emergence of adults in March. The adult females lay eggs in the midribs as well as in lateral axis of new leaves. Nymphs emerge from eggs during August-September and crawl to the adjacent buds to suck cell sap. As a result of feeding, the buds develop into hard conical green galls. The galls are usually seen during September-October. Consequently, there is no fruit set.

Control: The galls with nymphs inside should be collected and destroyed to prevent carryover of the pest. The pest can effectively be controlled by spraying of Monocrotophos (0.05%), Parathion (0.04%), Metasystox (0.1%) or Quinalphos (0.05%) at 2 week intervals starting from the middle of August. Repeated use of same chemical for every spray should be avoided.

Leaf Webber (Ortbaga euadrusalis): The infestation starts from the month of April and continues up to December. Eggs are laid singly or in clusters within silken webbings on leaves. Upon hatching, the caterpillars feed on leaf surface by scrapping. Later, they make web on tender shoots and leaves together and feed within. Old orchards with lesser space between tree canopy have more infestation than open orchards.

Control: Pruning of infested shoots and their burning in the month of April to July is found effective. Raking of the soil around the base of the trees in January, after the last generation has pupated, helps in checking the pest population. Three sprays starting from the last week of July at 15 days interval with Carbaryl (0.2%) or Monocrotophos (0.05%) or Quinalphos (0.05%) effectively controls the pest.

Stone Weevil (Sternochetus mangiferae): It is a common pest of mango in southern India. Varieties with high TSS and sugar such as Alphonso, Bangalora, Neelum, etc. are more prone to attack by this pest. Female lay eggs under the rind of ripening fruits. Newly emerged grubs bore through the pulp, feed on seed coat and later cause damage to cotyledons. Pupation takes place inside the seed. Pulp is discoloured around the affected portion.

Control: The pest population can be kept under check by destroying the affected fruits and exposing the hibernating weevils by digging the soil. Spraying the trees with Fenthion (0.01%) is found effective. In Alphonso and Banganapalli, a single spray of Monocrotophos 36 EC 1.5ml/litre of water at marble stage gave 100% and 97.5% control of stone weevil respectively. In Totapuri, Carbaryl 50 % WDP @ 4 g/ litre of water is effective in controlling stone weevil.

Chapter 3

Pest Control Tactics

The history of pest control probably began with the first human who ever swatted a mosquito or picked off a louse. From the fossil record, we know that all major taxa of biting flies and external parasites already existed by the time *Homo sapiens* first appeared on earth. *Phthirus* and *Pediculus*, the two genera of lice that feed on humans, have a host range that is limited to primates (apes and monkeys). And we suspect that human fleas (*Pulex irritans*) and bed bugs (*Cimex lectularius*) adopted cave-dwellers as hosts because these insects are most closely related to other species that live on bats. But since our primitive ancestors were hunters and gatherers, they probably found that insects were more useful as food than they were troublesome as pests. (Even today, people in some primitive cultures eat the lice they pick from one another's hair). It was probably not until the dawn of organized agriculture, when insects attacked the plants we grew for food, that we first recognized them as a potential threat to our own survival.

Pest control tactics were mentioned occasionally in writings of the ancient Chinese, Sumerian, and Egyptian scholars. Many of these tactics were embedded in religion or superstition, but a few had real scientific merit. Predatory ants, for example, were used in China as early as 1200 B.C. to protect citrus groves from caterpillars and wood boring beetles. Ropes or bamboo sticks tied between adjacent branches helped the ants move easily from place to place. A passage in Homer's Iliad (8th century B.C.) describes the use of fire to drive locusts into the sea, and the ancient Egyptians organized long lines of human drovers to repel swarms of invading locusts. Pythagorus, a Greek philosopher and mathematician, was credited with clearing malaria from a Sicilian town during the 6th century B.C. by instructing its residents to drain the marshes. Chemical substances that purportedly killed or repelled insects were in common usage. Many of these were of questionable value, but some worked, and a few are still in use today. Some of the inorganic compounds, such as sulfur and arsenic, have well-established insecticidal activity. And modern science has only recently come to recognize that many plant extracts used by ancient apothecaries (e.g, lemon oil, wormwood, hellebore, fleabane, etc.) do indeed contain compounds with useful activity against insects.

There was very little progress in pest control during the dark ages. Ignorance and superstition abounded. For what it was worth, St. Bernhard excommunicated the flies of his parish in 1121. In a book entitled "Natural History", Ferrante Imperato (1599) gave a prescription for eliminating flies from a dwelling:

> *"... draw the image of a fly... on a copper plate during the second half of the constellation of Pisces... then bury it in the centre of your house (during) the first half of the constellation of Taurus."*

With the Renaissance, people began to view insects less as a punishment from God and more as members of a natural world that could be studied and controlled. A renewed interest developed in insects both as organisms and as pests. More accurate observations of natural history and behaviour led to more inventive control practices. Hand labour was used extensively in early pest control, but cultural, physical, and chemical practices also evolved. Franz Ernst Brückmann, a German physician who lived in the early 1700's, designed the first mechanical traps for various insects. His fly traps consisted of a wooden box baited with a sweet attractant and equipped with a spring-loaded lid. When several flies had settled inside the trap, the lid was snapped shut to trap the flies inside. The volume of the trap was then decreased by sliding the ends of the box together and squashing the flies against their bait. Brückmann also designed flea traps. They were hollow, perforated cylinders, baited with blood or honey, and worn around the neck as a pendant. For a time, these flea traps, crafted from ivory or silver and often ornate in design, were popular fashion accessories worn by the aristocracy of western Europe.

Since the late 1800's, entomologists and chemists have made outstanding progress in the technology of pest control. Today's arsenal of weapons is large and diverse, encompassing legal, cultural, physical, genetic, and biological tactics, in addition to the well-known chemical pesticides. In general, all of these tactics work in at least one of the following ways:

- They kill the pest directly — usually by exposing it to lethal substances or unsuitable environmental conditions.
- They reduce the reproductive potential of a pest population — often by modifying its environment (biotic or abiotic) or by restricting its movement.
- They modify the pest's behaviour to make it less troublesome (attract it, repel it, confuse it, exclude it, mislead it).

Most of the control tactics that are commonly used today can be grouped into two broad categories: natural controls and artificial controls. By definition, a natural control may be any environmental factor that keeps a pest population below its economic injury level. Examples might include geographic barriers, cold temperatures, or natural enemies that keep population growth in check. Artificial controls, on the other hand, employ products or processes of human origin to modify a pest's distribution, behaviour or physiology.

Fly swatters and insecticides are two obvious examples. But as is often true in biology, there are some control strategies that seem to straddle the line, exhibiting characteristics of both groups or not fitting neatly into either category. In fact, some of our most effective tactics are natural controls that we can improve, enhance, accelerate, or augment by appropriate human intervention.

Cultural Control

Surprisingly simple modifications of a pest's environment or habitat often prove to be effective methods of pest control. As a group, these tactics are usually known as cultural control practices because they frequently involve variations of standard horticultural, silvicultural, or animal husbandry practices. Since these control tactics usually modify the relationships between a pest population and its natural environment, they are also known, less commonly, as ecological control methods. Simplicity and low cost are the primary advantages of cultural control tactics, and disadvantages are few as long as these tactics are compatible with a farmer's other management objectives (high yields, mechanization, etc.). Unfortunately, there are still a wide variety of insect pests that cannot be suppressed by cultural methods alone.

Crop rotation is one of the oldest and most effective cultural control strategies. Growing a single crop year after year in the same field gives pest populations sufficient time to become established and build up to damaging levels. Rotating the field to a different type of crop can break this cycle by starving pests that cannot adapt to a different host plant.

Farmers in the Midwest, for example, can reduce populations of wireworms (Elateridae) and rootworms (*Diabrotica* spp.) in corn fields by switching to oats, wheat, or legumes. Similarly, the clover root curculio (*Sitona hispidula*) that feeds exclusively on legumes can be eliminated by switching from clover or alfalfa to corn or small grains. Rotation schemes have also proven successful for controlling pests in pasture lands. Texas cattle fever, a protozoan disease transmitted by cattle ticks, *Boophilus annulatus*, has been eliminated from the southern United States due, in part, to a diligent campaign of pasture rotation designed to protect beef and dairy herds from tick infestation. Crop rotation schemes work because they increase the diversity of a pest's environment and create discontinuity in its food supply. As a rule, rotations are most likely to be practical and effective when they are used against pests that:

a. attack annual or biennial crops
b. have a relatively narrow host range
c. cannot move easily from one field to another, and
d. are present before the crop is planted.

Intercropping (also known as mixed cropping) is another way to reduce pest populations by increasing environmental diversity. In some cases, intercropping lowers the overall attractiveness of the environment, as when host and non-host plants are mixed together in a single planting. But in other cases, intercropping may concentrate the pest in a smaller, more manageable area so it can be controlled by some other tactic. Strips of alfalfa, for example, are sometimes interplanted with cotton as a trap crop for lygus bugs (Miridae). The alfalfa, which attracts lygus bugs more strongly than cotton, is usually treated with an insecticide to kill the bugs before they move into adjacent fields of cotton.

In some crops, it is possible to create discontinuity in the pest's food supply simply by altering the time of year for planting or harvesting. This strategy, often known as phenological asynchrony, allows farmers to manage their crop so it remains "out of phase" with pest populations. Sweet corn, for example, can escape most injury from corn earworms (*Helicoverpa zea*) if it is planted in early spring and harvested before larvae mature. In the Midwest, farmers delay planting winter wheat until after the "fly-free date" to protect their crop from injury by the Hessian fly (*Mayetiola destructor*).

These planting dates are calculated to ensure that wheat is not mature enough to attract egg-laying flies, yet still has enough warm weather to grow sufficiently before winter. Careful timing of harvest dates in alfalfa can cause high mortality in populations of alfalfa weevil (*Hypera postica*) and alfalfa caterpillar (*Colias eurytheme*) by removing edible foliage before the larvae have completed development. Harvest timing is often the most practical method available to foresters for controlling bark beetle (Scolytidae) infestations in pine plantations. Since trees become more susceptible to beetle outbreaks as they grow older, good management dictates that timber stands be harvested for lumber before the trees reach full maturity.

Managed application of water or fertilizer can have a big impact on the survival of pest populations in some crops. Annual flooding, for example, is a cultural practice that eliminates many potential pests in cranberry bogs. Regular rainfall (or overhead irrigation) can significantly reduce infestations of twospotted spider mites (*Tetranychus urticae*) in tree fruits. Good irrigation and appropriate fertilization keeps plants healthy, vigorous, and more resistant to insect injury. It is not unusual for small amounts of injury to actually stimulate compensatory growth in healthy plants.

Sanitation is another cultural control strategy that may be highly effective for some pests. Removing crop debris from cotton fields after harvest eliminates overwintering populations of pink bollworms (*Pectinophora gossypiella*), European corn borers (*Ostrinia nubilalis*), and sugarcane borers (*Diatraea saccharalis*). Collecting dropped fruit from beneath an apple tree reduces the next season's population of apple maggots (*Rhagoletis pomonella*), codling moths (*Cydia pomonella*), and plum curculio (*Conotrachelus nenuphar*). Shredding or burning the pruning wood from a peach orchard kills shothole borers (*Scolytus rugulosus*) and lesser peachtree borers (*Synanthedon pictipes*) that would otherwise emerge and reinfest the orchard.

Clean cultivation is often recommended as a way to eliminate shelter and/or overwintering sites for pest populations. Simply tilling or plowing a corn field before winter may disrupt a pest's life cycle by causing mechanical injury, by increasing exposure to lethal cold temperatures, by intensifying predation by birds or small mammals, or by burying the pests deep beneath the soil surface. Populations of corn earworms and European corn borers have been greatly reduced in recent years by community-wide efforts to plow under corn stubble after harvest.

Although it works well in corn and cotton, clean cultivation is not always the best solution for eliminating pest populations. In some cases, ground cover or crop debris shelters natural enemies that are important members of the agroecosystem. Without adequate shelter these beneficial populations would die or move away.

Biological Control

Natural control strategies that employ biological agents for pest suppression are generally classified as biological control tactics. In conventional usage, this term usually refers to the practice of rearing and releasing natural enemies: parasites, predators, or pathogens. A slightly broader definition of "biocontrol" includes any related management activity that is designed to protect or conserve natural enemies.

Biocontrol agents include a wide variety of life forms, including vertebrates, invertebrates, fungi, and microorganisms. These beneficial species are common in most natural communities and, although their presence is often unnoticed, they help maintain the "balance of nature" by regulating the density of their host or prey population. Insect species often become "pests" when this ecological balance is disrupted by natural events or human intervention. Biological pest control strives to reestablish this balance in one of three ways:

Importation. Foreign exploration is conducted to identify and collect natural enemies in the country from which an exotic pest has been introduced. Following the discovery of a potential biocontrol agent, it undergoes extensive evaluation to insure that its ecology and host range are compatible with the community to which it will be introduced and that it will not become a pest once it is released. Suitable candidates are reared and released in the new habitat in hopes that they will become established and suppress the pest population.

1. *Conservation.* A variety of management activities can be used to optimize the survival and/or effectiveness of natural enemies. Conservation activities might include reducing or eliminating insecticide applications to avoid killing natural enemies, staggering harvest dates in adjacent fields or rows to insure a constant supply of hosts (prey), or providing shelter, over-wintering sites, or alternative food sources to improve survival of beneficial species.
2. *Augmentation.* Natural enemies that are unable to survive and/or persist in a new environment can sometimes be reared in large numbers and periodically released to suppress a pest population. In some cases, small numbers of a beneficial species are released in several critical locations to suppress local pest outbreaks (an inoculative release). In other cases, larger

numbers are released in a single location to flood the pest population with natural enemies (an inundative release).

Examples of Biocontrol

Just because an insect is a predator or a parasite does not necessarily make it a good biological control agent. Since the early 1900's over 600 species of beneficial insects have been introduced into the United States. Of these, only about 20% have been outright successes as biological control agents; another 35% have been partially successful, and the other 45% never became established or failed to have any significant effect on pest populations. From these successes and failures, entomologists have concluded that introduced biocontrol agents are most effective when they exhibit the following characteristics:

Narrow host range. Generalized predators may be good natural enemies but they don't kill enough pests when other types of prey are also available.

1. *Climatic adaptability.* Natural enemies must be able to survive the extremes of temperature and humidity that they will encounter in the new habitat.
2. *Synchrony with host (prey) life cycle.* The predator or parasite should be present when the pest first emerges or appears.
3. *High reproductive potential.* Good biocontrol agents produce large numbers of offspring. Ideally, a parasite completes more than one generation during each generation of the pest.
4. *Efficient search ability.* In order to survive, effective natural enemies must be able to locate their host or prey even when it is scarce. In general, better search ability results in lower pest population densities.
5. *Short handling time.* Natural enemies that consume prey rapidly or lay eggs quickly have more time to locate and attack other members of the pest population. Small populations of efficient natural enemies may be more effective biocontrol agents than larger populations of less efficient species.
6. *Survival at low host (prey) density.* If a natural enemy is too efficient, it may eliminate its own food supply and then starve to death. The most effective biocontrol agents reduce a pest population below its economic threshold and then maintain it at this lower equilibrium level.

Just as predators and parasites may serve as effective biocontrol agents for insect pest populations, so insect herbivores have also proven effective as agents for controlling weed populations. Feeding injury by herbivores may weaken a plant or increase its susceptibility to disease, desiccation, freezing temperatures, or other herbivores. In any case, an injured plant is likely to be a poor competitor, and in time, beneficial species will have an opportunity to reclaim the vacated space.

Insect Weed-eaters

of the Klamath weed (*Hypericum perforatum*) and the prickly pear cactus (*Opuntia* spp.) are two well-known examples of successful weed control using insect herbivores.

Biological control is a particularly appealing pest control alternative because, unlike most other tactics, it does not always have to be reapplied each time a pest outbreak occurs. Once natural enemies are released into a new environment, there is a good chance they will become established and provide a self-perpetuating form of control. Biological control is also the only control tactic that increases, rather than decreases, the species diversity within an agroecosystem. This increased diversity often results in greater stability because wild fluctuations in population density are less common in communities with a diverse food web.

Biological control is not a "quick fix" for most pest problems. Natural enemies usually take longer to suppress a pest population than other forms of pest control and farmers often regard this as a disadvantage. It also may be difficult to "integrate" natural enemies into a crop or commodity when pesticides are still in use. Beneficial insects are often highly sensitive to pesticides and their resurgence (recovery to pre-spray densities) is usually much slower than that of pest populations. Rapid pest resurgence often leads to a vicious cycle of continued chemical usage that prevents natural enemies from ever becoming reestablished.

Perhaps the greatest potential for future progress in biological control lies in improving the success of microbial pathogens. Many of these organisms are highly desirable as biocontrol agents: they attack a narrow range of insect hosts, they are not hazardous to humans or domestic animals, and they do not pose a threat to the environment. As a group, though, pathogens have never proven to be quite as reliable as other forms of pest control. They are vulnerable to desiccation, ultraviolet radiation, and high temperatures. They may not survive long enough in the environment to encounter a host, and even if they survive, their virulence may be too low to overcome the host's defenses.

In the past few years, however, there has been encouraging progress in the development of microbial insecticides. These are commercial suspensions of spores, toxins, or virus particles that can be mixed with water and sprayed onto crops just like conventional insecticides. In many cases, microbial insecticides are better than conventional insecticides because they suppress pest populations without eliminating natural populations of predators and parasites.

A great deal of research is still needed before we can begin to benefit from the full potential of biocontrol. Fortunately, new developments in biotechnology may soon enable us to create new strains of microbial pathogens that are more virulent, easier to mass produce, and less sensitive to temperature and humidity. Work is currently underway to transform a common soil microbe, *Pseudomonas fluorescens,* into a pathogen of soil-dwelling insects by implanting the delta-endotoxin gene from *Bacillus thuringiensis*. Clearly, there are major changes in biocontrol that lie just over the horizon.

Breeding for Host Resistance

Breeding plants (or animals) for resistance to insects is really just another form of biological pest control. Rather than finding insects to attack the pests, breeders look for genetic traits (or combinations of traits) that reduce an organism's susceptibility to attack or injury by its insect pests. This idea was first tested in the 1870's by C. V. Riley, an entomologist who successfully fought a French outbreak of phylloxera (an aphid-like pest of grapes) with resistant North American rootstocks. Indeed, much of the pioneering work in this field has been done with crop plants because they are usually cheaper to breed and easier to manage than livestock. Recent innovations in genetic engineering, however, are likely to accelerate development of resistant genotypes in poultry, swine, and other commercial livestock.

In general, there are three approaches that plant breeders use to develop resistant cultivars:

1. *Antibiosis.* Plants produce a wide variety of defensive compounds (allelochemicals) that protect them from herbivores. These compounds may reduce growth, inhibit reproduction, alter physiology, delay maturation, or induce various physical or behavioral abnormalities in herbivores. By purposely selecting for plants with high levels of allelochemicals, or by breeding such plants with less resistant ones, it is often possible to develop new cultivars that resist pest injury yet still retain desirable horticultural characteristics.

2. *Antixenosis.* A physical or chemical property of a plant can make it so unpalatable that it is largely protected from herbivore attack. This type of resistance is often known as nonpreference. It may involve the presence of feeding repellents (or the absence of feeding attractants), or it may involve physical traits such as hairs, waxes, or a thick, tough epidermis that do not provide the pest with a desirable feeding substrate. Alfalfa, for example, has been bred with hairy leaves to deter feeding by the spotted alfalfa aphid.
3. *Tolerance.* Some plant genotypes are simply able to "tolerate" injurious insects better than others. Tolerant cultivars may be exposed to the same pest populations as susceptible ones, but they do not suffer as much injury. Many varieties of field corn, for example, have fairly narrow, brittle stalks. When attacked by European corn borers, these stalks are further weakened and break easily in a windstorm. This wind damage, known as lodging, makes the corn hard to pick with mechanical harvesters. Plant breeders have largely resolved this problem by breeding for corn with thicker, stronger cornstalks. These tolerant cultivars are still attacked by corn borers, but they "stand up" to the injury and insure a harvestable crop.

Developing resistant cultivars by traditional breeding methods can be a slow and uncertain process. But new advances in biotechnology now make it possible to pluck genes from one organism and insert them into the cells of a completely different species.

For example, the delta-endotoxin gene from *Bacillus thuringiensis*, an insect pathogen, has already been isolated and inserted into crop plants (notably cotton and tobacco). Cells of the genetically engineered plants can produce their own toxin, making them "resistant" to certain insect herbivores (e.g., Lepidoptera).

Although gene-splicing techniques are still in their infancy, they offer exciting new possibilities for expanding the diversity of resistant plant genotypes.

Legal and Regulatory Control

At first, it may seem ludicrous to even contemplate the use of legal measures as pest control strategies — it's easy enough to pass laws, but how do you train the insects to obey them? In reality, legal sanctions are aimed, not at the insects themselves, but at a range of human behaviors that are most likely to affect the dynamics of pest populations. Legal control tactics, therefore, include all forms of legislation and regulation that might prevent establishment or reduce spread of an insect population.

The Plant Quarantine Act of 1912 was the first legal action taken in the United States to prevent the introduction of pests from foreign countries. This law, and others that followed it, established a network of inspection stations at major ports of entry and gave the federal government authority to organize border quarantines, to inspect all agricultural products, and to restrict entry of any infested goods. Today, these inspections stations operate under the jurisdiction of the Animal and Plant Health Inspection Service (APHIS), a branch of the U. S. Department of Agriculture. APHIS operates about 85 inspection facilities, employs 400-500 inspectors, and intercepts over 20,000 potential pest introductions in an average year. It also operates the Plant Quarantine Training Centre where inspectors from all over the world are trained to spot problems before they cross international borders. These agents, now representing nearly 40 countries, cooperate in efforts to limit the spread of indigenous pests under an international agreement that requires inspection and certification of nearly all agricultural commodities prior to export.

Federal import/export laws have been enacted to establish standards for a wide range of agricultural products. Most of the fruits and vegetables imported to the U.S., for example, must receive some form of fumigation, heat treatment, controlled atmosphere storage, or radiation exposure before they are released from quarantine. These post-harvest treatments are designed to kill whatever pest species may be present without affecting taste or quality of the produce.

Despite the best efforts of APHIS inspectors, some pests still slip into the U. S. and become established (e.g., cereal leaf beetle), or move naturally across international borders (e.g., imported fire ants and Africanized honey bees). When this happens, APHIS cooperates with state and local governments to establish domestic quarantines for limiting the movement of these pests within our borders. Whenever appropriate, eradication programs may be launched together with quarantine actions in an effort to destroy the entire pest population while it is still small. Inspections and quarantines are also operated by some state and local government agencies. The California Border Patrol, for example, is notorious for its interceptions of uncertified agricultural products at the state line. The rigid laws that prohibit private citizens from bringing a bag of oranges into southern California are generally given credit for excluding numerous pest species that might have caused millions of dollars in losses. The price tag for these prevention efforts is only a fraction of the estimated $350 million Californians spent to clean up their most recent infestation of the Mediterranean fruit fly (*Ceratitis capitata*). Since 1915, nearly 300 pest species have been under some sort of domestic quarantine. The success of these programs has been erratic. In some cases (e.g., screwworm flies, khapra beetles, and parletoria date scale), eradication programs have managed to eliminate the pests. But for every noteworthy success, there has been a complete failure (witness the cereal leaf beetle, the gypsy moth, and the European corn borer). In general, eradication efforts are most successful when the pest population is relatively small, does not disperse rapidly, and is highly susceptible to control tactics.

Even if a newly established pest population cannot be eradicated, it can often be slowed down a bit by regulating the movement of its host plants or animals, prohibiting sales of infested products, and inspecting all shipments of affected goods. This type of containment strategy usually gives growers and producers a short reprieve before invasion of the pest and buys a little more time for researchers to find better methods of control. By using the containment approach in Texas during a 1971 outbreak of Venezuelan equine encephalitis, entomologists and veterinarians had enough time to reduce populations of mosquito vectors and immunize nearly 3 million horses before the disease could spread into other states.

Licensing and certification are important regulatory tools used in some commodities to ensure that infested or contaminated material is not sold commercially or used as breeding stock. Maintaining disease-free mother plants in strawberries, for example, can often prevent the spread of insect-borne disease during the growing season. Certified seed or plant material is produced under strict guidelines established by commodity organizations in their own interests. These products may be more expensive initially, but eventual savings in pest control usually provide a good return on the investment.

Eugenics and Birth Control

Some control tactics are designed to suppress a pest population by altering its genetic makeup and/or reducing its reproductive potential. As a group, these tactics are frequently known as genetic controls because they affect the accuracy or efficiency with which a pest species passes its genetic material (DNA) from one generation to the next. Genetic control usually works in one of two ways: either by causing (inducing) reproductive sterility, or by incorporating new and potentially deleterious genes (or alleles) into the genetic makeup of a pest population. In effect, some members of a pest species are transformed into biological time bombs that eventually destroy other members of their own species. Because of the self-destructive nature of these tactics, they are sometimes called autocidal control.

Insects can be sterilized by exposing them to certain chemical agents (chemosterilants) or to non-lethal levels of ionizing radiation (X-rays or gamma rays). Chemosterilants are really a form of chemical control. They usually work by blocking the onset of sexual maturity, by inhibiting the production of eggs and/or sperm, or by damaging the chromosomes. These compounds will be covered more fully in the section on semiochemical insecticides.

Exposure to radiation also damages chromosomes (usually by breakage or mutation). Since cells with damaged chromosomes cannot divide correctly, they do not form normal gametes or produce viable offspring. Although the susceptibility of each insect species is different, a proper dose of radiation administered at an appropriate stage of development (usually to pupae) can often induce sterility without causing other deleterious side effects. Sterile individuals, reared in large numbers and released into the environment, can mate with "normal" individuals but they produce no viable offspring. The more sterile individuals released, the fewer "normal" matings are likely to occur.

The effects of radiation on insect development was first studied in 1916, but this knowledge was not put to practical use until the 1930's when E. F. Knipling suggested that it might be possible to suppress pest populations by flooding the environment with large numbers of males that had been rendered infertile by irradiation. By maintaining a constant population of sterile males that was large in comparison to the number of virgin females, Knipling calculated that the number of "normal" matings would decrease each generation until the population was forced into extinction.

The first large-scale test of Knipling's sterile male technique was conducted in 1954 on the island of Curaçao against the screwworm, *Cochliomyia hominovorax*, a Dipteran pest of livestock. After just 13 weeks (four to five generations of the fly) and release of nearly a million sterile males, the screwworm population was completely eradicated from this tiny Caribbean island. The dramatic success of Knipling's first project led to larger screwworm eradication programs in the southeastern and southwestern United States and to intense interest in the sterile male technique for eradication of other pest species, including Mediterranean fruit flies (*Ceratitis capitata*), boll weevils (*Anthonomus grandis*), horn flies (*Haematobia irritans*), tsetse flies

(*Glossina* spp.), pink bollworms (*Pectinophora gossypiella*), and codling moths (*Cydia pomonella*).

Overall, the sterile male technique has had a spotty record of achievement. Despite isolated successes, no other eradication program has ever measured up to the outstanding performance of the screwworm project. From a theoretical standpoint, the use of sterile males is highly attractive because it is non-polluting, species specific, and becomes more effective as the target population decreases. But in practice, the technique simply doesn't work unless a pest population meets the following criteria:

1. *Easy to mass-produce.* A sterile insect release program cannot even be considered unless large numbers of insects can be produced at relatively low cost.
2. *Females mate only once.* This characteristic is uncommon among insects, but it is important for success of the sterile male technique because multiple matings only increase the probability that a female will receive sperm from a fertile male.
3. *Males can be sterilized without loss of competitive vigor.* Sterile males must be able to compete successfully for mates with normal (fertile) males. Sterile males must be physiclly and behaviorally identical to normal males.
4. *Low initial population.* The pest population must be small enough so the initial release of sterile males will outnumber the population of normal males. In some cases, alternative control tactics may be used to reduce initial populations to practical levels.
5. *Restricted geographic range.* Sterile males must disperse to all parts of the range. Small, localized populations are easier to control than large dispersed populations.

Within the last few years, geneticists have begun to devise new genetic control strategies that work by altering DNA or by adding new genes or alleles into the genetic makeup of a pest population. Most of these new tactics are still untested in field applications, but they do offer intriguing possibilities for the future. One such approach involves breeding members of a pest species that have been genetically altered to make them more susceptible to cold winter (or hot summer) temperatures. This type of genetic trait is known as a conditional lethal mutation; it causes mortality only when triggered by extreme environmental conditions. Genetically altered individuals are mass-reared, released into the environment, and allowed to breed with other members of their species. Hopefully, the mutation will spread throughout much of the pest population before the "carriers" are killed by cold (or hot) temperatures.

Another promising approach involves cytoplasmic incompatibility. Although insects lack an immune system, some species do have different "strains" that are not reproductively compatible: the egg cytoplasm of one strain may contain substances that block or inhibit sperm from another strain. By mass-rearing and releasing large numbers of these "incompatible" insects, it may someday be possible to eradicate a pest population and replace it with a more "benign" strain.

Chemical Control

Most people understand that chemical pest control involves the use of chemical substances to kill or disrupt the life cycle of an insect pest. But few people outside the circle of entomology realize just how diverse these compounds are, and in how many different ways they can be used. Although conventional insecticides, the poisons, are still a mainstay of chemical control, they are gradually being superseded by less toxic compounds that disrupt insect development or modify behaviour. Some of these new chemical weapons are much safer for the environment and more species specific than most conventional insecticides.

Semiochemicals — Chemical Control of Behaviour

Much of an insect's behaviour is mediated by chemicals in its environment. By turning these chemicals to our own advantage, it is often possible to attract pests to traps or baits, or repel them from our homes, our crops, or our domestic animals. Behavioral messages are delivered by a wide array of chemical compounds. As a group, these compounds are known as semiochemicals. In some cases, they may facilitate communication between the members of a single species (e.g., pheromones) or between members of different species (e.g., allelochemicals). Functionally, semiochemicals may have a wide range of activity. They may serve as attractants or repellents, they may stimulate or inhibit feeding, they may provoke flight or inhibit it, or they may simply elicit behaviour patterns at inappropriate times.

Attractant pheromones and allelochemicals can be used as lures or baits in a wide variety of insect traps, or they can be mixed together with toxicants to produce an "elixir of death". Protein hydrolysates, for example, serve as feeding attractants for fruit flies (*Rhagoletis* spp.). These chemicals can be applied to sticky traps to improve catch, or combined with an insecticide and sprayed on fruit crops to suppress active infestations. Phenethyl propanoate, eugenol, and geraniol can be mixed in a 3.5:3.5:3 ratio and used as an attractant for Japanese beetles (*Popillia japonica*). These are the active ingredients in the "floral attractant" found in popular bag traps for Japanese beetles. In some cases, chemists have produced synthetic compounds that are even more attractive than naturally occurring chemicals.

Trimedlure, a synthetic substitute for alpha-copaene, is produced commercially as an attractant for the Mediterranean fruit fly (*Ceratitis capitata*). Improved food lures and baits are among the most promising new developments for controlling cockroaches (Blattoidea) in homes and businesses. These are the active ingredients in a new generation of "roach motels" where the insects "check in but don't check out."

Sex pheromones are among the most powerful of chemical attractants. Ever since they were first discovered by A. A. Budenandt in 1959 (from silkworm moths, *Bombyx mori*), these chemicals have aroused great interest because of their potential as pest control agents. During the past 30 years, chemists have identified the sex pheromones for over 300 insect species. Many of these compounds are now sold commercially. In some cases, pheromones are packaged (or encapsulated) in slow-release dispensers (rubber septa, hollow fibres, or rope wicks) that are used as lures in traps of various designs. At low densities, these pheromone traps are a valuable monitoring tool, providing information on the density and distribution of pest populations. At high densities, they can be used for mass trapping sexually active adults (usually males) in efforts to reduce population density and lower a pest's reproductive potential.

Slow-release formulations of sex pheromones can also be used for mating disruption. By increasing the concentration of pheromone in an insect's environment, it may be possible to make everything smell like a prospective mate. Males wear themselves out courting inanimate objects or become habituated to the odor and stop responding to it. This approach, variously known as air permeation or the innundation technique, has shown promise for controlling a number of fruit and vegetable pests, including the codling moth (*Cydia pomonella*), the cabbage looper (*Trichoplusia ni*), the oriental fruit moth (*Grapholita molesta*), and the peachtree borer (*Synanthedon exitiosa*).

Chemical repellents and feeding deterrents are also useful tools for managing insect behaviour. As their name suggests, these compounds cause insects (or other arthropods) to disperse or to discontinue normal feeding behaviour. Repellents such as dimethyl phthalate, benzyl benzoate, and N,N-diethyl-m-toluamide (DEET) have been developed to protect humans from biting flies, ticks, and chiggers (immature trombiculid mites). Other compounds, like di-n-butyl succinate or butoxypolypropylene glycol, are used as fly repellents for cattle. Moth balls and flakes (paradicholorobenzene or alpha-naphthalene) are placed in drawers and closets to prevent infestation by a variety of insects that feed on stored products and natural fibres. Woolen cloth can be manufactured with colourless dyes (e.g., Eulans and Mitin FF) that bond permanently to the fabric and make it unpalatable to clothes moths and carpet beetles. Wood preservatives, such as pentachlorophenol, act as feeding deterrents to termites and other wood-dwelling insects.

The neem tree, *Azadirachta indica* (Meliaceae) is a promising new source of feeding repellents that may be developed for use on selected non-crop plants. The leaves, twigs, and seeds of this tree, which is grown commercially in India, contain at least 25 biologically active compounds that act as insect repellents, feeding deterrents, or growth regulators. Azadirachtin, the most abundant of these active ingredients, is now commercially available in the United States. Sold under the trade name Margosan-O, this new feeding deterrent has been approved for use on non-food greenhouse crops, ornamentals, and turf.

Chemosterilants — Chemical Control of Reproduction

There are over four hundred chemical substances that are known to cause reproductive sterility in insects. Some of these compounds inhibit ovarian growth and development, while others appear to induce fundamental changes in the chemical structure of nucleic acids (DNA and RNA). These changes (mutations) prevent cell division or obstruct normal embryonic development. Chemosterilants belong to several major chemical groups. These compounds are applied directly to the insect or incorporated into food that serves as a bait.

All chemosterilants are extremely hazardous compounds. Their effects are not restricted to insects; they also cause cancer, birth defects, and other mutations in humans and domestic animals. Clearly, these chemicals cannot be dispersed in the environment like other pesticides. Instead, they must be applied under controlled laboratory conditions, usually to insects that are mass reared and released as part of a sterile release program. Although there is much interest in finding a chemosterilant whose effects are limited to insects, no such compound has yet been found.

Insect Growth Regulators — Chemical Control of Development

The enzymes and hormones that regulate developmental processes within an insect's body can sometimes be exploited as chemical control weapons. These compounds, often known as insect growth regulators (IGRs), can be used to stimulate development at inappropriate times or inhibit it at other times. The major groups of IGR compounds include:

1. *Chitin inhibitors.* These chemicals (e.g., diflubenzuron and teflubenzuron) inhibit the molting process (apolysis) by blocking the activity of chitin synthetase, an enzyme needed by epidermal cells when constructing a new exoskeleton. Because of this mode of action, chitin inhibitors are highly specific to arthropods. They act rather slowly (2-5 days), but eventually disrupt any process that involves construction of new cuticle (e.g., molting, hatching, pupation). They are most effective when used against the immature stages of a pest. Diflubenzuron, currently registered under the trade name Dimilin, is used for controlling gypsy moths, boll weevils, and various other pests.
2. *Molting Hormone Analogues.* Ecdysteroids stimulate the molting process by mimicking the action of molting hormone. Applied to the surface of an insect's body or incorporated into its food, these compounds work by initiating premature ecdysis during the immature stages of development. Ecdysteroid-like compounds have been found in some plants where they evidently serve as a defense against insect herbivores. But despite their potential as insect growth regulators, the ecdysteroids have never been developed into commercial products. Their chemical structural is similar to that of human reproductive hormones (estrogen, progesterone, and testosterone), and like many other steroid compounds, they have the potential to cause cancer and birth defects.
3. *Juvenile Hormone Analogues.* Juvenile hormone (JH) and related compounds act as insect growth regulators by inhibiting the developmental changes associated with embryogenesis, morphogenesis, and reproduction. During normal development, JH levels are elevated in larvae (or nymphs) and decrease prior to pupation (or adult eclosion). Contact exposure to JH analogues during the egg stage or after the last larval (or nymphal) molt can inhibit development, delay maturation, and eventually result in death. Since the onset of mortality is usually quite slow (days to weeks), JH analogues have limited utility in agriculture. But several compounds (e.g., hydroprene, kinoprene, and methoprene) have been successfully incorporated into household products for controlling ants, fleas, and other household pests.

4. *Anti-juvenile Hormones.* These unique compounds (the precocenes) were first isolated in 1976 from a common houseplant (*Aegeratum houstonianum*). Precocenes are cytotoxic agents. They become activated by enzymes in the insect's corpora allata, selectively destroying these glands, and preventing all subsequent production of juvenile hormone. In immature insects, exposure to anti-JH compounds may result in premature (precocious) development of adult structures or behaviors. In adults, precocenes can cause sterility because the presence of juvenile hormone is necessary for normal production of eggs and sperm. Anti-JH compounds seem to be most effective against Hemipterans. Despite their unique mode of action, these IGRs have never been developed into commercial products because they break down too rapidly in the presence of oxygen.

Conventional Insecticides — The Killer Chemicals

Conventional insecticides are among the most popular chemical control agents because they are readily available, rapid acting, and highly reliable. A single application may control several different pest species and usually forms a persistent residue that continues to kill insects for hours or even days after application. Because of their convenience and effectiveness, insecticides quickly became standard practice for pest control during the 1960's and 1970's.

Overuse, misuse, and abuse of these chemicals have led to widespread criticism of chemical control and, in a few cases, resulted in long-term environmental consequences. The effectiveness of an insecticide usually depends on when and where the pest encounters it. Most insecticides are absorbed directly through an insect's exoskeleton. These compounds are known as contact poisons because they are effective "on contact". Other insecticides act as fumigants. They are released in the vapour state (as gases) and enter the insect's body through its tracheal system. Fumigants are most effective when they are used in an enclosed area such as a greenhouse, a warehouse, or a grain bin. Still other compounds must be ingested before they have an effect. These are known as stomach poisons. They often work more slowly than fumigants or contact poisons, but they are still useful for certain types of pest control in homes and businesses.

Systemic insecticides are a special type of stomach poison. These compounds are absorbed by the tissues of a plant (or animal) without ill effects. Insect pests ingest the insecticide when they feed on the treated organism. Systemic insecticides are sometimes included in the diets of domestic animals to protect them from internal parasites (e.g., cattle grubs and other bot flies). Plant systemics can be incorporated into the soil around ornamentals or bedding plants. The insecticides are absorbed by the roots and translocated to leaves, stems, and flowers. If the insect that feeds on a treated plant doesn't acquire a lethal dose of insecticide, it may at least be deterred from further feeding. Although systemic insecticides are commonly applied to horticultural plantings, they are not as useful for many food crops because the insecticide remains in the food after harvest.

Insecticides contain one or more active ingredients that serve as toxicants (poisons). In their purest form (technical grade), these chemicals may be too toxic, too unstable, or too volatile to be handled or applied safely. Therefore, technical grade insecticide is always mixed with other compounds, known as adjuvants, in order to improve the performance, safety, or handling characteristics of a commercial product.

These mixtures (technical grade insecticide plus adjuvants) are known as formulations. Almost anything could be an adjuvant: pumice, ground walnut shells, buffalo gourd root powder, vegetable oil, etc. These compounds are usually listed on the label as "inert ingredients", but they are certainly not inactive. Many adjuvants are proprietary products, protected by patents and closely guarded as industrial secrets. They may represent 90-95% of the total volume of a commercial formulation. An insecticide can be manufactured in a variety of formulations, each tailored for specific applications. Some of the more common formulations include:

Granules (G) or pellets (P) — Coarse particles (e.g., clay, ground corn cobs, or walnut shells) can serve as a carrier for certain types of insecticides. The toxicant slowly leaches out of the carrier, minimizing its movement within the ecosystem and maximizing its active life (persistence). Granular formulations are commonly used for controlling soil-dwelling insects and in systemic insecticides that are applied around the base of plants. Since granular formulations do not blow or drift, they are considered relatively safe from the standpoint of accidental human exposure.

Dusts (D) — These dry powders are usually formulated with inert particles of ash, chalk, talc, or clay. They are designed to be sprinkled or blown onto target surfaces. The powder sticks to feet, legs, and other body parts of passing insects, and the toxicant is eventually ingested when the insect cleans itself. Insecticides that act as stomach poisons are often formulated as dusts. These formulations are not always suitable for outdoor applications because they have a tendency to drift (blow away with the wind).

Soluble powders (SP) or wettable powders (WP) — Dry formulations that are mixed in water to form homogeneous spray solutions. A soluble powder dissolves completely in water, whereas a wettable powder contains an emulsifier that produces a uniform suspension (colloid). Many foliar insecticides are formulated as wettable (or soluble) powders because they are easy to transport and generally have a long shelf life.

Emulsifiable concentrates (EC) — These liquid formulations contain the toxicant(s) and an emulsifier dissolved in organic solvent. The concentrate is diluted with a large volume of water to produce the final spray mixture. Some foliar insecticides are formulated as emulsifiable concentrates. Unlike most wettable powders, they do not leave a visible residue on fruits and vegetables. Sensitive plants, however, may be injured by organic solvents in the mixture. Emulsifiable concentrates are also commonly used in sprays for urban and industrial pests.

Aerosols (A) — Insecticides that are formulated together with a solvent may be pre-packaged in pressurized spray cans or sold unpressurized for use in special fogging machines. Spray cans are relatively expensive (per pound of active ingredient) but they are convenient, easy to store, and have a long shelf life. Commercial foggers are typically used indoors (e.g., greenhouses and warehouses) or for control of biting flies in community-wide pest control operations.

Ultralow-Volume Concentrates (ULV) — These highly concentrated formulations (more than 8 pounds of active ingredient per gallon) are designed to be used in specialized spray equipment that atomizes the concentrate droplets. ULV spray equipment is used by most aerial applicators (airplane sprayers) who treat forested lands or large agricultural acreages.

Despite their many advantages, conventional insecticides are not ideal pest control agents. Indeed, one of their greatest strengths, broad-spectrum activity, is also one of their greatest weaknesses. While it is certainly an advantage to control multiple pest species with a single chemical treatment, the non-specificity of most conventional insecticides poses a serious threat to non-target organisms in the environment. High mortality among natural enemies can have an enduring impact on the ecological balance of any community. In the absence of biocontrol agents, more insecticide applications may be the only recourse available to stop pest resurgence. Once we step onto this "insecticide treadmill", it can be very difficult to get off.

The non-target effects of insecticides are not limited to natural enemies. Most of the organochlorine compounds (e.g., DDT) have been banned from use in the United States because of their indirect effects on reproduction in birds of prey (e.g., eagles, ospreys, condors, etc.). These environmentally stable compounds are highly soluble in lipids, making it possible for them to accumulate in the body fat of non-target organisms. Predators, particularly those at the top of a food chain, amass pesticide concentrations many times greater than anywhere else in the environment. This process, known as bioaccumulation (or biomagnification), was responsible for high levels of DDT and related compounds in birds of prey during the 1960's and 1970's. These pesticides did not injure the birds themselves, but caused thinning of their egg shells and high rates of breakage during incubation. Most organochlorine insecticides were banned during the 1970's and 1980's and many of the threatened bird species are now recovering from the brink of extinction.

In some cases, sub-lethal concentrations of an insecticide can stimulate rather than suppress the growth of a pest population. This phenomenon, known as hormoligosis, has been observed in a number of pest species, including twospotted spider mites (*Tetranychus urticae*), western corn rootworms (*Diabrotica virgifera*), and brown planthoppers (*Nilaparvata lugens*). Low doses of pesticide seem to improve the nutritional quality of host plants, thereby increasing a pest's reproductive potential or decreasing its time of development.

And finally, there is the problem of resistance to insecticides. Insects are among the most adaptable organisms on the face of the earth. For the past 400 million years they have managed to survive by adjusting to changes in their environment, so it should come as no surprise that they can also adapt to chemical pesticides. Resistance has increased exponentially since the the late 1940's, and today, there are over 500 pest species that exhibit some level of resistance to at least one type of insecticide.

Insects may become resistant to insecticides in several ways. Biochemical resistance usually involves changes in the metabolic pathways that insects normally use to break down plant defenses and other environmental toxins. This detoxication is facilitated by enzymes (esterases, hydrolases, transferases, and oxidases) that change the chemical structure of toxicants before they cause physical harm. Physiological resistance involves functional changes in basic life processes that alter the way toxicants interact with the body. Some German cockroaches (*Blatella germanica*), for example, have become resistant to carbaryl (a carbamate insecticide) as a result of genetic changes in the permeability of their cuticle. Behavioral resistance may occur as a result of any innate change in behaviour that reduces an insect's probability of encountering a toxicant. In some parts of Panama, the mosquitoes that vector malaria (*Anopheles albimanus*) have become hypersensitive to certain insecticides. They manage to escape lethal doses of insecticide by avoiding enclosed areas and refusing to land on treated surfaces.

When an insect population becomes resistant to one insecticide, it may also become less susceptible to other toxicants in the same chemical family. This phenomenon, known as class resistance, is a common problem in all major groups of insecticides (organochlorines, organophosphates, carbamates, and synthetic pyrethroids). When an insecticide loses its effectiveness because of pest resistance, users typically replace it with another compound from a different chemical group. But resistance to one group of compounds does not prevent subsequent development of resistance to compounds in other chemical groups.

Over the years, some pests have developed multiple resistance to nearly every insecticide that has ever been used to fight them. Colorado potato beetles, *Leptinotarsa decemlineata*, on Long Island, N.Y. are probably the best example of a population that has developed multiple resistance. It may never be possible to prevent insects from developing resistance, but there are ways to manage these chemicals that will prolong their useful life.

Integrated Control

Just as ground, air, and naval forces are integrated to achieve military objectives, the tactical weapons of pest control can also be integrated to achieve more effective management of pest populations. Development of resistance, effects on non-target organisms, and damage to the environment can all be minimized with selective and judicious use of multi-faceted control tactics. This approach, commonly known as integrated control, requires an understanding of ecological principles as well as a thorough knowledge of the pest's life history and population dynamics.

Integrated pest control is not a new concept. It was commonly practiced in the years before synthetic organic insecticides became widely available. But the old ways were largely abandoned after World

War II because chemical weapons were so effective, convenient, and inexpensive. Once we recognized the dangers of over-dependence on a single control strategy, the principles of integrated pest control gained renewed acceptance.

Today, integrated pest control forms the foundation of Integrated Pest Management programs (IPM) that take a comprehensive and multi-disciplinary approach to solving pest problems. Insects, weeds, plant diseases, and even some vertebrate pests (e.g., birds and rodents) are included under the IPM umbrella. These programs emphasize management rather than eradication. They take a broad ecological approach to pest problems, focusing on all members of a pest complex in an effort to identify the optimum combination of control tactics that will reduce pest populations below economic thresholds and maintain these levels with the least possible impact on the rest of the environment. This approach, often called biorational pest control, relies heavily on cultural and biological tactics that are supplemented with carefully timed applications of highly selective chemical weapons.

The complexity of modern IPM programs will continue to increase as we add more knowledge about pests, develop new management tactics, and learn how to optimize existing control strategies. Mathematical models of population growth rates and plant-pest interactions are being developed and incorporated into computer programs that help us assimilate and interpret data from the many different variables that affect pest population dynamics. These computer programs, usually known as expert systems, are yet another tool in our expanding stockpile of pest control technology.

Integrated pest management is also a major component of Sustainable Agriculture, a holistic approach to modern farming that encompasses not only pest control, but also conservation of soil and water resources, and a wide range of other management practices (e.g., composting, mulching, green manure, waste disposal, etc.). The goal of sustainable agriculture is to satisfy human needs for food and fiber while preserving nonrenewable resources, protecting environmental quality for future generations, and safeguarding the profitability and long-term viability of commercial agriculture.

Chapter 4

Cultural Control Practices

Cultural Controls: Making Your Site Unattractive to Pests

Cultural control methods include properly selecting and rotating crops, sanitizing and solarizing the soil, choosing the best planting and harvest times, using resistant varieties and certified plants, taking advantage of allelopathy, and intercrop-ping.

Crop Rotation

Certain pests are more common in some crops than in others. Rotating crops to different sites can isolate pests form their food source or can change the conditions pests must tolerate. If another site is not available, change the type of crops grown in the garden plot. Do not put members of the same plant family in the same location in consecutive seasons.

For example, do not follow melons with cucumbers or squash.

This is also true for rotations using green manure crops, which add organic matter to the soil when they are tiled in before they produce flowers or seeds. Waiting two years to plant the same family of vegetable in the same location is the most effective rotation practice; however, yearly rotations can also be beneficial. Rotating annual flower plantings is also a good practice.

Sanitation

Many organisms responsible for disease and insect problems overwinter in plant debris such as shriveled fruit. Diseases on these shriveled fruit infect new leaves following spring. Removing crop residues, weeds, thatch, and volunteer.

Soil Solarization

A clear plastic sheet spread over the soil traps solar heat, which kills soilborne diseases, insects, nematodes, and many weed seeds. The treatment should occur during summer's high air temperatures and intense solar radiation. Keep the soil damp during the solarization process, and keep the plastic in place for several weeks. OSU Extension Fact Sheet EPP-7640 explains soil solarization in more detail.

Timed Plantings and Harvests

Many crops may be planted or harvested early to miss heavy pest infestations, while still achieving a full yield. Plant-ing earlier than normal may involve the use of cold frames or hot caps to protect seedlings form the weather while they get a head start growing. The crop then has a competitive edge over pests. Early planting depends upon the gardener knowing the emergence times and life cycles of the pests to be controlled.

Resistant Varieties

When buying seeds or plants, try to choose those with built-in resistance to diseases and nematodes. Sources for this information include OSU Extension Fact Sheets, seed catalogs, and plant and seed packages. It may be better to forego some production capability in favor of the increased pest resistance, if you must make such a choice. During the growing season, stressed plants can lose their resistance to pests, so be sure the crop has the water and nutrients it needs. When shopping for seeds and plants, check the labels for abbreviations similar to these, used to designate various types of pest resistance or tolerance:

A—Alternaria stem canker

ALS—angular leaf spot

ANTH—anthracnose

CMV—cucumber mosaic virus

DM—downey mildew

F—Fusarium (race 1)

FF—Fusarium (races 1 & 2)

MDM—maize dwarf mosaic

N—nematode

NCLB—northern corn leaf blight

PM—powdery mildew

SCLB—southern corn leaf blight

St—Stemphylium (gray leaf spot)

SW—Stewart's wilt

TLS—target leaf spot

TMV—tobacco mosaic virus

V—Verticillium.

Certified Plants

When they are available, consider buying plants labelled as "certified" or grown and inspected under sterile or quarantined conditions. Certified plants may cost more than others, but the certification guarantees they are free of diseases. Strawberries and potatoes are among crops which may be offered as certified plants.

Allelopathy

Allelopathy, a natural chemical interaction among plants, has been the subject of much research recently. Allelopathy refers to stimulatory as well as inhibitory properties. A living plant may release toxins, or in the case of decaying plant tissues, microorganisms may play a role in the release of the toxin. The microbes may also modify nontoxic compounds into toxic compounds. Black walnut trees and Johnson grass are among plants that have been shown to inhibit the growth of winter annual weeds and may offer some control of root knot nematode.

Intercropping or "Companion Planting"

The premise of companion planting is that certain plants repel insects, or attract beneficials that attack the insects. There is no significant data to prove the value of companion planting or intercropping, but it is thought that certain plants may produce substances which confuse insects, altering their impact as a pest. Some evidence also shows that planting flowers among vegetables attracts beneficial wasps seeking the flowers' nectar, and those wasps lay their eggs in the larva of certain pest species.

There is a popular but largely inaccurate belief among gardeners that marigold will control nematodes and other insects if planted among vegetables. Most marigold varieties do not have this capability. Only the French marigold (*Tagetes patula*) varieties, such as Nemagold.

Vegetables Families

Tomato Family: tomato, potato, pepper, eggplant.

Onion Family: onion, shallot, leek, chive, garlic.

Beet Family: beet, Swiss chard, spinach.

Cole Crop Family: cabbage, cauliflower, broccoli, brussels sprouts, bok choy, collards, kale, kohlrabi, mustard, radish, rutabaga, turnip.

Legume Family: bean, pea, cowpea, peanut.

Carrot Family: carrot, celery, celeriac, parsley.

Cucurbit Family: cucumber, watermelon, cantaloupe, pumpkin, squash, gourds.

Lettuce Family: lettuce, chicory, endive.

Green Manure Crop Family: hybrid sudangrass, buckwheat, soybean, cowpea, mung bean, garden pea, fava bean, ryegrass, rye grain, barley, oats, vetch, Austrian winter pea, clovers, greens.

Cultural Control of Drinking by Universal Moderation or Differential Access

The prospects for a new generation of cross-cultural hypotheses and analyses of drinking behaviour quietly improved a good deal over the past decade because of the emergence of a new and still largely untapped store of data. These data come from a growing series of broadly comparable social surveys focused on drinking behaviour and carried out in a variety of cultural settings. The trend began in 1970-1979 with, first, a four-country Scandinavian drinking survey carried out in Iceland, Norway, Sweden, and Finland (Makela, 1984, 1986) and, second, a three-culture study coordinated by the World Health Organization and carried out in community-level samples in Mexico, Scotland, and Zambia (Rootman, 1983).

The WHO study—called the "WHO Community Response to Alcohol-Related Problems" project — spawned two additional comparable and nearly contemporaneous surveys in Canada (Smart, 1980) and in the U.S. (Roizen, 1981). More recently, surveys incorporating at least part of the WHO questionnaire have been done in Spain (Caetano and Martinez, 1987), in Japan (Clark, 1986), in a second region of Mexico (Caetano and Medina-Mora, 1986), and among Spanish-surname subgroups in the U.S. general population (Caetano, 1986).

An August 1986 conference in Washington, D.C. — which drew participants from researchers involved in the original WHO project, from researchers at work on subsequent WHO-like projects, and from others interested in this emerging resource— bespeaks the continuing vitality of this new tradition of research.

These studies offer a tonic to cross-cultural research not only because they are new and unfamiliar but because survey data bring a new kind of information to the territory, thus forcing new modes of thinking upon us. And although these survey data are not as richly textured as ethnographic reports nor as comprehensive in their coverage of the world's cultures as the Human Relations Area Files, they nevertheless offer the field a valuable new window on cross-cultural variation and a methodology that nicely complements the more familiar brands of cross-cultural data.

For example, survey data almost invariably require an observer/ analyst interested in cross-cultural variation to pay equal heed to variation *within* each cultural setting as well. Cross-cultural hypotheses trying to make their way in this sort of data environment must somehow manage to incorporate both cross-cultural and within-cultural variation into their explanatory agenda — no small duty.

Today, I would like to illustrate some of the charms of this sort of data by drawing on WHO-study data from the three original settings and the U.S. data, which in fact derive from a survey of Contra Costa County California in 1979. I have two goals. One is to illustrate how ordinary survey data can be used to examine structural and normative issues, issues that are usually reserved for other kinds of data and analysis styles.

The other is to offer one or two conjectures on cross-cultural theory borrowing from these same data. Before we begin, however, a word or two should be said about the data sources. My survey data comes from seven separate samples: one in California, one representing the Lothian (Edinburgh) region of Scotland, two representing two kinds of communities — one urban and the other rural — in Mexico, and, finally, three samples representing an urban residential, periurban, and rural settlement in Zambia.

Technically speaking, it is problematic to combine the samples drawn within a given country into a single sample, but that has been done nevertheless. Similarly, you should be aware that these data do not reflect the national populations of these settings but merely specific communities therein. Data, moreover, have been weighted everywhere besides the U.S. in order to bring the samples into equality across the genders.

Cultural Permission to Drink

Who respondents in Scotland, Mexico, Zambia, and the U.S. were asked to estimate much drinking would be "all right" for males and for females of four specific, hypothetical or test ages, 16, 21, 40, and 60. Four pre-coded answers were offered:

(1) no drinking,
(2) one or two drinks only,
(3) enough to feel the effects but not drunk, and
(4) "getting drunk is sometimes alright."

The data I will be presenting are the frequency distributions for the whole samples in each setting; they tell us not the answers of males or of females regarding these age/sex statuses but instead the answers of the whole population about males and about females.

To begin, let us consider the normative patterns relating to the cultural control of access to drinking in general, a question we can address by initially collapsing together all three pro-drinking response categories. The Scottish and U.S. "permission" patterns were very similar. Both showed little difference in permission rates between the sexes (though females ran slightly below males), and both suggested an age-plateau form with high— and even nearly universal — permission granted for the three test ages twenty-one and over. Simple as they are, these curves seem to hint at three kinds of connections between drinking norms and the wider cultural environment — these links appearing as cultural premiums (1) on majority age, (2) on universalism or egalitarianism, and (3) on legalism.

Concerning majority age, the normative pattern we see is consistent with a high cultural premium on a single, consensual, and officially-sanctioned "threshold" age of majority, one governing the right to drink as well as other adult prerogatives. The flatness of the curves across the three "adult" ages and the similarity of the curves for the sexes suggests a high cultural premium on universalism or egalitarianism.

Finally, I see legalism hinted in the close agreement between popular opinion and official laws governing the right to drink. Legalism, in turn, suggests that the state holds and articulates the issue of cultural approval for drinking in the Scottish and U.S. settings.

The Mexican and Zambian permission patterns were quite different both from each other and from the universalistic and age-plateaued pattern of the Scottish/U.S. data. The Mexican data showed a substantial divide in cultural permission across the sexes. The Zambian data, on the other hand, showed wide variation in permission across test ages. Notice, for example, that all four female age-statuses in the Mexican sample are accorded lower rates of cultural permission than three of the four male age-statuses.

We have ordered the Zambian results from highest to lowest cultural permission level (first 40-year-olds, next 60-, next 21-, and, finally, 16-year-olds). Notice in this chart that the results follow a clear-cut pattern of age stratification: 40-year-old males and females ranked 1st and 2nd respectively in cultural permission levels; 60-year-old males and females ranked 3rd and 4th; 21-year-old males and females ranked 5th and 6th; and, finally, 16-year-old males and females ranked next-to-last and last.

To be sure, no matter what the age-status, Zambian males were accorded higher levels of permission than females (even considerably higher for 21-year-olds), but more importantly, females in each age-status group were consistently accorded a permission level at least as high as that accorded males in the next-lower test-age group.

Permission levels in the Zambian case, then, can fairly be said to be stratified primarily by age whereas permission in the Mexican case was stratified primarily by sex. In both the Mexican and the Zambian cases, moreover, permission to drink appears to involve a *more gradual or graded* cultural license in place of the *sharply divided* pattern at majority-age suggestion in the Scottish and U.S. permission norms. The Mexican and Zambian normative patterns suggested neither the majority-age premium, nor the universalism, nor the legalism hinted in the Scottish/U.S. pattern.

On the contrary, in both Mexico and Zambia the data reveal that the right to drink is much more strongly tied to one's ascribed status, or, in other words, to bases of social stratification more characteristic of traditional societies. Mexican and Zambian permission norms, then, seem to have 'stained-out' (to use Duster's [1983] evocative term) the ascribed-status bases of each society's broader social stratification systems, much as a biologist's dye stains tissue on a microscope slide. Mexico's gender-based and Zambia's age-based stratification systems are not in and of themselves particularly surprising.

Mexico, of course, is well known for its machismo men and modest women, as Africa is known as home to many age-based systems of social stratification. Yet the striking difference between the two bases of stratification — gender and age — raise interesting questions. What are the chief differences between ascription based on age and on gender (La Fontaine, 1978)? How is it that permission-to-drink norms reflect the underlying architecture ascribed stratification in each cultural setting? What does the apparent parallelism between drinking permission and stratification tell us, on the one hand, about drinking and, on the other, about stratification? What are the chief differences in drinking practices, attitudes, and problems between universalistic and ascriptive social structures?

Finer-grain puzzles are also suggested in these findings. What does it mean, for example, that in Mexico *21-year-olds* (of either sex) were granted higher permission levels than like-sexed *60-year-olds*, whereas in Zambia the reverse was true — *60-year-olds* were granted higher levels 21-year-olds? Did the gradual, non-plateaued form of Mexican and Zambian permission norms imply that adulthood itself is likewise defined in an incremental or cumulative way? Why did the norms applying to 60-year-olds decline in levels of permission from those of 40-year-olds in Mexico and in Zambia as well as (if only in a hinted-at form) in the Scottish and U.S. data?

Permission to Drink and the Power of the State

If the permission-to-drink normative patterns hinted at different levels of involvement by the state in the regulation of drinking, then just such a relationship was unexpectedly lent support elsewhere in the WHO data, once more nicely illustrating the potential for survey data to address structural societal variables. As the study's title suggests, the WHO researchers were interested in community responses to alcohol-related problems. Its survey questions in this territory happened to include a series of four vignettes about hypothetical alcohol-related dilemmas. This series provided data on the perceived severity of alcohol-related problems in the community, on perceptions of such problems' perceived prevalences, and — most relevant here—on preferred kinds of social control responses. (Data on only three of the four vignettes will be considered here.) The first vignette described a man who had drunk so much that "he falls down on the road and cannot get up" (MANFALL); the second described a man who has on two occasions struck his wife when he was drunk (HITWIFE); the third described a man "who spends so much money on drinking that there is not enough food for his family" (NOFOOD).

Included in the battery of questions asked about each vignette were three questions asking respondents to evaluate the appropriateness of the involvement of various people or agencies in helping or controlling the man in question. For example, the questions asked (1) whether the man's relatives ought to help him, (2) whether a bystander on the street should do something, and (3) whether the police or appropriate authorities should be called in.

This series was introduced into the survey questionnaire in order to monitor norms and attitudes regarding the social handling of one or another sort of drinking-related problem. Yet its results also painted a clear picture of the relative ascendancy of the differential involvement of the state — represented by the "police/authorities" involvement — across the Scottish, Mexican, and Zambian samples. Thus, it is worth briefly examining the different Patterns of control reported in these three setting.

In the Zambian samples roughly 90 percent agreed that relatives should help in all three vignette situations. Concerning bystander involvement, 90 percent supported bystander involvement for the MANFALL situation and from 75-80 percent supported it for the HITWIFE and NOFOOD situations. Less than half the samples supported the involvement of the police or appropriate authorities in the MANFALL and HITWIFE situations, but almost three-quarters supported such involvement in the NOFOOD situation. Involvement by police/authorities ranked third — behind relatives and bystanders — for all three situations. The Scottish results were quite different.

There, the vignette situations were seen as highly differentiated and sorted into two groups — the MANFALL situation, on the one hand, and the HITWIFE and NOFOOD situations, on the other. For all situations, involvement by relatives was supported by 80-85 percent of the sample. A like percent supported bystander intervention in the MANFALL situation, but involvement by police/authorities in this situation was supported by less than half the sample. The HITWIFE and NOFOOD situations — both of which involve third-party victims of the central actors drinking — were not seen as appropriate occasions for the involvement of bystanders by roughly three-quarters of the sample; in these, police/authorities was approved by over three-quarters of the sample. The patterns suggest an alternativity relationship between bystander and police/authorities involvements and a clear division of responsibility between the benevolent interventions (of a bystander) and the potentially punitive interventions (of police/ authorities) — with state involvement contraindicating involvement by the citizen on the street. Clearly, the state's monopoly of police power and the differentiation and definition of its role is much greater in the Scottish than in the Zambian case. Interestingly, the Mexican pattern of responses appeared to fall between the Scottish and Zambian patterns.

Charts 14D, E, & F present the same data as 14A, B, & C this time contrasting each cultural setting with respect to a single vignette situation. Chart 14D reports the MANFALL situation, in which all three cultural settings reveal quite similar response patterns — high on relative and bystander involvement and low on police/authorities involvement. In the HITWIFE situation, however, Chart 14E throws the issue of state control in the Scottish setting into sharp relief. Here we see that the Zambian response pattern to HITWIFE is not much different from the response pattern to MANFALL. The Scottish pattern shows a clear preference for the involvement of police/authorities and only low support for bystander involvement. In the Mexican samples bystander and police/authorities involvements are roughly equally supported by about two-thirds of the sample. Finally, in the NOFOOD situation, the Zambian pattern shows more support for police/authorities involvement, but such involvement appears to be added to high support levels for involvement by bystanders rather than substituting for it. In the Scottish and (though less so) in the Mexican case, as noted already, involvement by police/authorities is perceived more as an alternative to bystander involvement. These patterns, then, give us a clear if only cursory glimpse of the normative expectations surrounding relatives, citizens on the street, and civil authorities or police in regard to exemplary problematic situations. They illustrate the marked difference in the expectations regarding state control across the three cultural settings examined and lend a measure of support to the idea that the growth of the state's power is probably structurally associated with the universalistic normative patterns evidenced in the earlier permission-norm data we examined.

Cultural Permission To Drink Enough to Feel Alcohol's Effects

Though the Scottish and U.S. norms governing access to drinking suggested a universalist form, it is well known, of course, that even in these settings drinking behaviour is strongly associated with (or marked by) gender and age variables. Indeed, the WHO data from this same study reaffirmed this long-standing truth. For example, fully 80-85 percent of Scottish and U.S. young males (18-29) reported drinking at least as often as once a week; whereas among older females (50+) the proportion drinking that frequently dropped to only 25-35 percent. Concerning such relationships, the draft WHO monograph noted that "when we examine frequent drinking, in all four cultural settings the distributions for males of all ages fall entirely above the distributions of their female compatriots." What light might the normative data shed the normative arrangements for heavier drinking? Let us return data from the age/sex drinking norms series, this time with that series' responses dichotomized so as to divide respondents who would permit "feeling the effects" or sometimes getting drunk" to a hypothetical status from those who suggested "no drinking" or only one or two drinks maximum. This dichotomy will tell us more about the normative structuring of higher-quantity drinking in each setting.

The Scottish and American patterns are interesting both for their similarities and their differences. First, it is apparent in both that permission to drink is not equivalent to permission to drink enough to feel alcohol's effects. In both sites, too, the permission distributions now taken on more shape — that is to say, deviated from the "flat plateau" age-structure evidenced in the earlier permission-to-drink curves.

This suggests more of a culturally sanctioned age-based hierarchy with respect to higher-quantity drinking. In both settings and for both sexes, moreover, the apparent age-hierarchies evidenced in this new shape are the same: first, 40-year-olds, next 21-year-olds, next 60-year-olds, and, finally, 16-year-olds — this, a status with very low approval for higher-consumption drinking. Interestingly, the age-based hierarchies for both Scottish and U.S. females reveal a wider gap in permission levels between 40-year-olds and 21-year-olds than occurs between 21-year-olds and 60-year-olds. The same relative spacing is in evidence for Scottish males, if more weakly, but disappears for U.S. males.

Perhaps the most striking result of the Scottish and U.S. comparison, however, is the higher permission levels enjoyed by Scottish males in contrast with either Scottish females, U.S. males, or U.S. females. To take a More than 60 percent of the Scottish sample approved feeling alcohol's effects for a 40-year-old male, whereas the comparable approval levels for 40-year-old Scottish females was a little less than 50 percent, and for U.S. 40-year-old males and females roughly 40 percent. These results imply that the hierarchy of higher-consumption permission in the Scottish sample is ordered both by sex and by age whereas in the U.S. data — though small gender differences permission rates are evident — the permission hierarchy is geared to age alone.

The Mexican and Zambian data reveal different patterns. In the Mexican sites, permission to feel the effects is very low for females of all four ages. Indeed, females are by and large simply not permitted to drink enough to feel alcohol's effects — the highest approval level was won by 40-year-old females and that amounted to only about 15 percent. For approval ran from about 30 percent, for 21-and 60-year-olds to as much as 45 percent for 40-year-old males. Notice that although 21-year-old males received a ten-percentage point higher approval level than 60-year-olds for drinking, per se, approval levels for feeling the effects were roughly equal for the two ages. Sixteen-year-old males received approval from only about 5 percent of the sample, a figure that appears to be quite low until it is contrasted with the approval levels granted females.

Zambia's pattern is different yet again. The striking finding here is the high levels of approval granted "feeling the effects" for 40- and 60-year-old males and females. On balance, the data for Mexico, Scotland, and the U.S. have shown that approval levels for feeling the effects among the three statuses 21-years or older run about half as high as approval levels for any drinking — a little higher for Scots, a little lower for the U.S. and Mexican samples. In the Zambian case, however, the ratio of approving effects to approving drinking jumps to roughly 70 percent. In the Zambian case, then, approval of drinking is almost equivalent to approval of feeling alcohol's effects — suggesting the possibility of quite different structural arrangements for drinking.

Some Speculations

It is intriguing to contrast the U.S. and the Zambian patterns with respect to permission-to-drink and higher-consumption permission. The U.S. norms suggested, of course, a pattern that permitted drinking universally to all adult statuses but imposed a normative obligation to restrict drinking to moderate levels. In the Zambian case permission to drink is a differentially allocated but the norms on consumption are much weaker. I have called these two models of normative control "universal moderation" (UM) and "differential access" (DA) — and it will be noted that they provided the title of my presentation today. My hunch is that these two models of normative control may carry broad implications for the structuring of drinking behaviour and offer an interesting line of connection between drinking behaviour, on the one hand, and other dimensions or aspects of culture.

The responses we have examined refer to "ideal" norms rather than actual conduct. We have seen, however, that even ideal norms in UM normative settings like the U.S. and Scotland pertain more to permission to drink per se than to heavier drinking. Drinking per se, then, appears to be symbolically associated with the universalistically defined rights and responsibilities of adulthood in these settings, but heavier drinking is symbolically associated with a separate stratum of cultural meaning, and one drawing on more traditional status arrangements.

I was particularly impress, for example, by the close agreement in heavier-consumption norms between U.S. and Mexican males. This result may, of course, be pure coincidence, but it nevertheless highlights the interesting possibility that drinking norms, particularly in less traditional cultures, occupy a point of tension or stress between two competing normative systems. The difference between the U.S. and Scottish heavier-consumption norms is also interesting. If the structure of drinking's norms were tied firmly to the level of modernization in a society, then we would generally expect Scottish and U.S. norms at both permission levels to be the same. The standard indicators of modernization theory are, after all, quite broad and would give us no reason, I suspect, to call either the U.S. or Scotland the more modernized of the two. It follows, then, that the observed differences in U.S. and Scottish heavier-consumption norms suggests that modernization-type variables are not the only dimension of culture at work in shaping drinking's normative structure.

This, I think, is a telling possibility and may open up an interesting analytical approach to teasing out of survey data which aspects of culture, after all, are tied most securely to drinking. Just such findings, as it happens, can prove helpful in interpreting some of the other data collected in such surveys. For example, indicators of "alcohol dependence" in the Scottish and U.S. samples revealed strikingly similar frequencies in the two samples across six of the seven items used. The seventh item, however, showed U.S. males and females reporting much higher frequencies than the Scots. This item, as it happens, asked if one had felt that he or she should stop drinking or cut down in the past year — a measure, perhaps, of both self-recrimination and the "dryness" of drinking norms. Could it be that the more permissive normative structure observed for Scottish men explains this curious difference in symptom frequencies. But if so, why do Scottish females differ from U.S. females on this measure as much or more than Scottish males differed from U.S. males?

Weed Management

Lesson Introduction

Explain to students that a great deal is already known about noxious weed control. One of the aspects of control is that a multi-pronged attack is usually taken with chemical, biological, and cultural control methods all being used either simultaneously or in sequence. Much research centres around development of plant specific controls, cost of controls, and reducing "collateral damage" that is inherent in the control methods utilized. Be sure to explain that the cost of fuel is always a consideration, whether it is fuel to run the spray rig and tractor, fuel to deliver the goats or bugs to the site and haul away the weeds you pull from the ground. This section of lessons has the opportunity to be expanded greatly. The two CD-Rom disks are very detailed and each section of each disk gives students a chance to study a different aspect of weed management. If students acquire current values for products and services, they can gain a greater understanding of the impact that weeds have on ranching, especially on cattle production.

Montana Education Content Standards & Benchmarks

Vocabulary

Acre – *noun*, 43,200 square feet or 1/640 of a square mile (5280' × 5280').

AUM – *noun*, animal unit month, amount of grass one cow eats in one month (about 600 pounds).

Biological control – *noun*, the type of living organism used to control the number of weeds, such as flies, sheep, and fungi.

Chemical control – *noun*, a process of controlling the number of weeds by spraying chemical herbicides on the plants. These chemicals are usually specific for broadleaf plants.

Cultural control – *noun*, a process of controlling the number and spread of weeds through plowing, pulling, mowing, burning, or irrigating.

Entomologist – *noun*, a scientist who studies insects and their life cycles.

Fungus – *noun*, a plant that lacks chlorophyll and obtains its nutrients from a host organism.

Gall fly – *noun*, a biological control that reduces the number of seeds produced by knapweed plants. Gall flies lay eggs in knapweed seed heads.

Herbicide – *noun*, a chemical used to control, suppress, or kill weeds by severely interrupting their growth process.

Infestation – *noun*, a troublesome or hostile spreading that threatens to overrun or inhabit an area in such large numbers that other organisms could be harmed.

IPM – *noun*, Integrated Pest Management, a program for weed control that uses all the available methods of control, rather than using just a single method.

Irrigate – *verb*, supply water to an ecosystem or crop land artificially to increase the productivity of the land.

Larva – *noun*, the immature, wingless, and often worm-like feeding form that hatches from the egg of many insects.

Section – *noun*, 640 acres or one mile X one mile.

Student Activities

1. Read and understand the "Montana Agricultural Statistics" pdf files. Research current costs and values of items mentioned and insert those present values and costs in place of those given in the handout.
2. Read the "Joe Smith Ranch Information Sheet" and calculate the results of the problems shown in questions 1-7 from. Show and label all work.
3. Read the "Chemical Weed Control Information Sheet" and answer questions 1-6. Show and label all work.

Cultural Control Strategies

Mulching

Mulches of wood chips, straw, grass clippings, paper, rocks, rubber, black plastic, or landscape fabric can be used to control weeds. Mulching is typically done on relatively small areas where the application can be monitored. Mulching works best on annual weeds and cannot control some perennial weeds because their extensive food reserves allow them to continue to grow up through the mulch. Wood chips are often used in forestry/wildland settings where woody weeds and undesirable vegetation can be shredded into wood chips and left onsite as mulch.

In order to be effective as a weed barrier, the mulch should be 3 – 6 inches thick. Landscape fabrics are also used as mulches but are more commonly used on a smaller scale as they tend to be more expensive. These fabrics are heavy, sometimes woven and are made of polypropylene or other weather-resistant materials. Water passes through them slowly and many are made to prevent weeds from growing through them. When using organic mulches like wood chips, straw, or grass make sure the material comes from a weed free source and/or does not contain any weed seeds or fragments.

Burning

The most effective fires for controlling invasive plant species are typically those administered just before flower or seed set, or at the young seedling/sapling stage. Sometimes prescribed burns that were not originally designed to suppress an invasive species have that happy side effect. But in some cases, prescribed burns can unexpectedly promote an invasive, such as when their seeds are specially adapted to fire, or when they resprout vigorously. These prescriptions must be modified or other management actions taken to undo or reverse the promotion of the invader.

Most successful weed control efforts that result from burning are due to the restoration of historical (natural) fire regimes, which had been disrupted by land use changes, urban development, fire breaks, or fire suppression practices. Many prescribed burn programs are, in fact, designed to reduce the abundance of certain native woody species that spread into unburned pinelands, savannas, bogs, prairies, and other grasslands. Repeated burns are sometimes necessary to effectively control weedy plants, and herbicide treatments may be required to kill the flush of seedlings that germinate following a burn.

Spot-burning invasive weeds with a propane torch can be cheaper and easier than implementing a prescribed fire (permits are still required), but is only effective when the infestation is small. Spot-burning can be used to burn individual plants, groups of plants in a small area, or to ignite brush piles. Propane torches can be used in areas where there is little or no fine fuel to carry a prescribed burn, and can also be used to kill plants when conditions are wet.

Prescribed Burning and Herbicides

Some invasive species have underground storage organs that resprout vigorously after fire, and/or seeds whose germination is stimulated by fire. Some of these species may not be possible to control with fire, but some can be controlled with repeated burns and others may be especially vulnerable to herbicides after a burn. Resprouts or seedlings that are 1 to 3 months old are often especially sensitive to herbicides.

Timing of Burn

The timing of a burn can strongly affect the fire's impact on native and exotic plant populations. For example, in California's Carrizo Plain Natural Area, Meyer & Schiffman (1999) determined that warm-season prescribed burning (late-spring and fall) was most effective for reducing abundance of Mediterranean annual grasses. Native plant cover and diversity also increased significantly following warm-season prescribed burns. Winter burns, however, did not affect the abundance of native plants, and exotic plant cover was only moderately reduced. Timing was also key in controlling smooth brome *(Bromus inermis)* and encouraging the growth of native grasses in Nebraska and Minnesota (Willson & Stubbendieck 2000). Timing prescribed burns so that they occurred at the time of tiller (aboveground lateral stem) elongation, yielded an immediate and persistent reduction in both tiller density and biomass of smooth brome.

Burning in Extensively Disturbed Areas

Not all burn treatments in wildlands are beneficial. When fires become too intense, crown-fires and death of native plants that typically survive fires can result. If temperatures are too hot, soil organisms and seeds, even those of species that require fire stratification for germination, may perish, and valuable soil nutrients may be volatilized or otherwise lost. In extensively disturbed areas of southwest Australia, fire actually enhanced the invasion of weeds along roadsides, and resulted in an overall decrease in the abundance of native species (Milberg & Lamont 1995). Schwartz & Heim (1996) reported that fire was at best moderately successful for garlic mustard (*Alliaria petiolata*) control in Illinois forests, and Luken & Shea (2000) determined that repeated prescribed burning had no significant effect on garlic mustard in Kentucky. In both cases, however, the burns were detrimental to native herbaceous species, reducing both density and richness. Even three years after the initial burns, native plant composition did not recover to pre-burn values.

Preventing Spread of Weeds

Keep all equipment, trucks, and engines clean of weed seeds. After each burn, and before moving to another site, be sure to clean (hose-off) all equipment, tools, and clothing used. This will minimize changes of carrying weed seeds directly to a new site where a fire might provide perfect conditions for their establishment.

Cultural Control of Established Dandelion in Turf

Dandelion can be a major weed problem in turf. It forms dense clumps which affects the footing and the overall playability of sports fields and golf courses. The texture and colour of its leaves does not blend harmoniously with the surrounding grasses. When in bloom, its yellow flowers are yet a further detractor from the aesthetics of the turf. Dandelion seed is spread on the wind, and as this seed can be carried great distances, prevention of new infestations is difficult. Once a dandelion becomes established and supported by its large taproot, they are even harder to control. The foundation of cultural dandelion control in turf has always been to create a growing environment in which the turf has the competitive vigor to repel the establishment of new dandelion seedlings. However, when it comes to dealing with established dandelion, many of the cultural methods are abandoned, opting to solely rely on chemical control. As concerns over the use of herbicides for aesthetic purposes increases, the turf manager must explore new strategies and rethink the whole dandelion control process.

In a previous study conducted by the Prairie Turfgrass Research Centre, mowing various species of grasses at a higher height was thought to reduce weed populations. A number of grasses resisted the development of weeds when mowed at 3" and there was virtually no weed development over the duration of the study. Aeration and topdressing are cultural practices that improve gas diffusion in soil which in turn helps to promote root growth. In addition, aeration physically removes thatch from the turf which also helps to improve the turf quality. High quality turf as promoted by adequate fertility has also shown to reduce weed infestations. In addition, the two organic fertilizers, soybean and corn gluten meal, have shown to have weed suppression properties. This study was developed to determine the effects of various cultural strategies for the control of dandelion in turf.

Methodology

The effect of three cultural turf management practices on an established dandelion population was tested at the Prairie Turfgrass Research Centre on a site located at Olds College, Olds, Alberta. Two weeks prior to the establishment of the trial (July 8), a pretrial treatment of the herbicide Trillion was applied to the site in an attempt to suppress the dandelion infestation. Seven days after the herbicide had been applied the level of control was visually assessed. On June 22 nd the trial was established over the pretreated area and the location of each plot was identified and staked to allow for future reference. Test plots that measured 1x 1 metres were laid out on a turf site with a uniform stand of established dandelion. Each treatment was replicated four times within a three factor randomized complete block design. To determine if soil aerification was a factor to consider when controlling established dandelions, half of the plots were aerified, while the other half was left untreated. A Ryan GA30 aerator, equipped with 5/8"hollow tines and set for a 3"x 3" spacing, was used to effectively open up the turf. The soil cores were harvested and removed from the site. USGA specification sand was spread over the aerified turf. Several passes with a drag mat were made to fill the core holes and to evenly distribute the sand over the aerated area.

The second factor evaluated the effect of fertility on weed control. The individual fertilizers were weighed into plot sized lots and applied by hand using a simple shaker bottle. The plots received the initial application of fertilizer on the establishment date and a second application of the fertilizer was applied four weeks later. The mowing treatments were conducted on a weekly basis. This process was accomplished by first mowing the entire experimental area at the higher cutting height of 2½" (6.25cm), then using a push mower, selectively recut specific plots at the lower height of ¾" (1.9cm).

On August 31 st the trial was evaluated for the first time. The dandelions in each of the one square metre plots were physically counted. In addition the plots were visually evaluated for overall turf quality. Following National Turfgrass Evaluation Program (NTEP) protocols, three turf quality factors: colour, density and area cover were assessed. The colour factor subjectively evaluated the uniformity and intensity of the colour displayed by the turf.

A 1 to 9 scale was used to rate each plot. Treatments which stimulated a uniform dark green colour received scores ranging from 6 for an acceptable colour to 9 for turf with outstanding colour. Plots that produced a lighter turf colour were scored lower. Density, the second quality factor, was subjectively evaluated for the impact of the treatments on the turf to produce more shoots and tillers. The 1 to 9 scale was again used to rate each plot. Treatments which stimulated a tightly knit turf received scores ranging from 6 for an acceptable density to 9 for superior turf. Treatments associated with a weak or thin turf stand were scored lower.

The final quality factor, area cover, subjectively evaluated the vigor of turf. Once again a 1 to 9 scale was used to rate each plot. Treatments which stimulated a thick competitive turf cover received scores ranging from 6 for an acceptable area cover to 9 for a superior area cover. Treatments producing a weak turf, affected by weed encroachment and/or the presence of bare patches, were scored lower.

Results

Herbicide Pretreatment

Seven days after the herbicide had been applied the level of control was visually assessed. Classic signs of herbicide damage were evident throughout the trial site as the top growth of the dandelions appeared twisted and some upper leaves were beginning to turn brown. At this point, good control was expected. However, as dandelions counts ranged from 2229 dandelions per m 2 on the August 29 rating date, effective control was questioned.

Mowing Height

An analysis of the data indicates that there was no statistical difference in the dandelion population when the two mowing heights were compared. Even though the turf mowed at the higher height scored better than the turf mowed at the lower height for colour, density and area cover, it was not statistically different than the turf that was mowed at a lower height.

Turf Aeration

At the ninety percent confidence level the turf which was cored and top dressed had a significantly higher number of dandelions per square metre than the nonaerated turf. The turf colour and turf density of the nonaerated turf was significantly better from that of the aerated turf. There were no significant differences between the treatments for area cover.

Fertilizer

At the ninety percent confidence level the high rate of the controlled release synthetic fertilizer had fewer dandelions per square metre than the other fertilizer treatments. The turf colour stimulated by the high rates of the corn gluten and the controlled release synthetic was significantly better than the turf colour of the unfertilized treatment and the lower rates of either the soybean meal or the controlled release synthetic fertilizer.

Discussion

In the initial year of this study, the infestation of dandelion had remained very high. The pretrial herbicide treatment provided only short term suppression of the dandelion's top growth. With only a single set of data points it is still unclear as to which treatments or combination of treatments will have the greatest effect on the dandelion population.

Cultural Control of Herbicide-resistant Lollum Rigidum Gaud. Populations in Winter Cereal in Northeastern Spain

Lolium rigidum Gaud. is one of the most common weed species in winter cereals in Northeastern Spain. Herbicide resistance of *Lolium rigidum* Gaud. to chlortoluron and diclofop-methyl has been reported in Catalonia since 1995 (Taberner *et al.*, 1995). The problem has continued growing and more populations have been found resistant to tralkoxydim, fluazifop and chlorsulfuron (Taberner, 2005). The estimated affected area is still less than 0.5% of the 6 million hectares of winter cereal in Spain (Díaz and Gorrochategui, 2005) but the affected fields require costly management. Low cereal yields and selling prices have encouraged Spanish farmers to invest minimum efforts in their cereal crops spraying expensive graminicides only when the cheap ones failed. However, the current high cereal price situation may introduce new elements in this scenario modifying the prospects for herbicide resistance.

The great capacity of *L. rigidum* for developing resistance adapting to many different compounds has led to the development of multiple resistant populations in Australia (Powles, 1993) and to the existence of biotypes resistant to herbicides belonging to 10 different mode of action groups (Heap, 2005). Taking into account that climatic and edaphic conditions of some regions in Australia and Spain are similar, the herbicide resistance problem can potentially grow in Spain. In order to prevent and control this problem it is necessary to develop non-chemical control strategies, adapted to the local farmers' practices (Matthews, 1996; Gill and Holmes, 1997; Nalewaja, 1999).

Out of these options, probably the easiest to be adopted in the Spanish context are delayed sowing, increasing weed density and ploughing. Stubble burning and crop topping are not authorised, pastures are not sown, and the resistance problem is probably not serious enough to justify modifying the external combine harvesters for seed catching as proposed by Matthews *et al.* (2004). If the problem grows, as it occurred in Australia, Spanish farmers will probably also adopt more non-chemical management strategies, improving these methods even years after the outcome of the problem, as exemplified with the seed burner presented by Matthews *et al.* (2004), thus showing that there is still interest in offering new alternatives to be combined with herbicides.

Two main aspects of the seed biology of *L. rigidum* offer opportunities for its control. First, it is known that the viability of the seeds in the soil is short (Gramshaw and Stern, 1977; Taberner *et al.*, 1992) so that burying them by ploughing can reduce their viability rapidly. For seeds buried at 0, 5 or 20 cm depth, Taberner *et al.* (1992) found that one year afterwards only around 5% of the seeds remained viable in Spanish conditions. This behaviour opens the possibility of burying the seeds with a deep ploughing.

A second important characteristic of *L. rigidum* is that main germination occurs after rainfall in the beginning of the season (Gramshaw and Stern, 1977). Cheam *et al.* (1998) observed main emergence in the two weeks after crop sowing and up to 93% emergence concentrated in the 5-6 first weeks, while Matthews (1996) (using data of Heap, 1988) describes how 60-80% of the *L. rigidum* seeds emerged in the initial germination flush in Australian conditions. This fact offers the opportunity of delaying the sowing date controlling the first *L. rigidum* emergence flush before sowing.

However, depending on the climatic conditions, germination can also be staggered. In some cases, in Spanish conditions, there is a concentrated germination flush during the first 40 days after cereal sowing (Planes *et al.*, 1999) but in other cases the germination period can start 2-4 weeks after sowing and last 4-7 weeks (Recasens and Kuc, 1998). Sowing delay aiming a reduction in weed populations can thus be very variable, depending on the climatic conditions of that year. An important inconvenient of sowing delay is the possible yield decrease (Planes *et al.*, 1999; Anderson, 2007).

Lolium rigidum control caused by increased sowing density has been scarcely reported. Gill and Holmes (1997) stated that Australian farmers are usually not willing to increase wheat plant density above 100-150 plants m-2. In Central Spain, Lacasta *et al.* (2004) tested the effect of sowing barley at three different densities (80, 160 and 240 kg ha-1) during eleven years in a monocrop situation finding that the intermediate rate (160 kg ha-1) generally yielded the most. In irrigated winter cereal, Moreno *et al.* (2002) recommended using more than 100 kg ha-1 in barley, as they did not find any significant yield increase in any of the three years after testing five different rates up to 200 kg ha-1 in the semiarid Spain. In different experiments at variable *L. rigidum* density, Izquierdo *et al.* (2003) did not find significant differences in barley yield for densities between 75 and 300 kg ha-1.

Liebman and Gallandt (1997) recommended combining several control techniques to have additive 148 *A. Cirujeda and A. Taberner /* Span J Agric Res (2009) 7(1), 146-154 effects, the so called 'many little hammers'. Similarly, Anderson (2007) proposed the dualistic approach of prevention and control, which is apparently effective in the semiarid steppe of the United States. However, little work on these techniques have been conducted in our area.

The objectives of this work were to test the effect of sowing delay, ploughing, increased cereal rate and the combination of these methods on *L. rigidum* control in winter cereal. The final aim was to provide new data on the effects of these methods, which should be combined in an Integrated Weed Management Programme to control herbicide resistant *L. rigidum*.

Material and Methods

Plant Material

Field trials were conducted on different herbicideresistant populations at different locations of the Lleida province. All experiments were conducted on commercial fields using the varieties chosen by the farmer and applying conventional fertilization. The resistance to herbicides was checked before treatment using a seed-based quick-test and confirmed in herbicide field trials. All the populations were resistant to chlortoluron, diclofop-methyl and tralkoxydim at the same time, excepting the population of barley in Sarroca, which was only resistant to chlortoluron. *Lolium rigidum* plant density in the control plots was high in all cases. Plots were placed perpendicular to the straw swaths, where this weed emerges predominantly (Blanco-Moreno *et al.*, 2004) so that *L. rigidum* distribution was similar between plots.

Cultural Control Methods

Sowing delay was conducted at five fields; mouldboard ploughing at four fields; the combination of both practices at two fields; increasing the cereal density and increasing the cereal density combined with sowing delay at one field. The experimental design was a split-plot with ploughing type as a main factor at Ferran 1, Verdú and Concabella. At Santa Maria and Ferran 2, the treatments were arranged as a randomised complete block design with two factors. At Sarroca, the experiment was a randomised complete block design with one factor. Elementary plots measured 4 x 5 m in all cases.

Ploughing was conducted with a five-furrowmouldboard plough at Ferran 1, Concabella and Verdú at 15, 20 and 25 cm depths, respectively. At Ferran 2 a four-furrow plough was used inverting the soil approximately at 18 cm depth. Commercial ploughs were used in all cases. At Santa Maria, normal sowing density was 175 kg ha-1 corresponding to 439 seeds m-2 and high sowing density was 1.5 fold the normal rate, that is, 250 kg ha-1 corresponding to 627 seeds m-2. Normal densities in the area were used in the rest of locations.

Assessments

Lolium rigidum density was assessed at weed and crop heading (BBCH scale 56-59, BBCH Working Group, 1997) to describe the final effect on the weed population. Assessments included number of *L. rigidum* plants, number of *L. rigidum* ears and total fresh biomass in 0.1 m2 squares. Biomass was determined immediately after cutting in the field. Crop plant density per linear meter was assessed three times in each plot at heading at Santa Maria (28th May 2003), Verdú (17th June 2003) and at Sarroca (27th April 2004) to describe the effect of the cultural methods on the crop.

Data Analysis

Data were transformed when necessary to satisfy normality and variance homogeneity following the indications of the Box-Cox Transformation (Bowley, 1999) using $(x + 1)0.5$ for most cases, excepting weed weight per plant at Ferran 1 where $x0.5$ was used. Analyses of variance were performed using the SAS GLM procedure (SAS Institute, 1991) for each location separately considering the different models. Least significant differences (LSD) were calculated for pairwise comparisons. Mean values are shown as least square means (LSM) and differences between means were analysed on the basis of ordinary t-tests considering $P < 0.05$ in the case of significant interactions.

Results

Mouldboard Ploughing

Mouldboard ploughing reduced significantly *L. rigidum* density at Concabella and Verdú. This difference was particularly remarkable at Concabella, where ploughing reduced 96% weed density. Although ploughing reduced 55% weed density at Ferran 1, this difference was not significant. Similar results were obtained for weed biomass in these three sites. No significant differences in densities and biomass were observed at Ferran 2. The lack of efficacy of ploughing at this site is probably due to the high stoniness at this location, which impeded deep enough ploughing without damaging the tool. Ploughing did not affect wheat density at Verdú, the only site where this parameter was measured.

Sowing Density

Increasing sowing density had no effect on crop density and on *L. rigidum* density and biomass at Santa Maria, the only site where this comparison was conducted.

The Combination of Different Techniques

The combination of sowing delay with ploughing or with sowing density did not improve the effect of the individual treatments significantly at any of the three sites but had some effect at Ferran 1, where *L. rigidum* density was reduced. At Ferran 1 and Verdú, ploughing was the most important control method. At Santa Maria, the main weed reduction was due to sowing delay. Concerning the effect of these techniques on the crop, at Santa Maria, increasing the sowing rate at the second sowing date gave significantly higher wheat density than the other treatments. However, no effect was observed at Verdú on the crop due to the combination of both techniques.

Discussion

Efficacy on weed plants due to ploughing was not considerable at Ferran 2, around 50% at Ferran 1, 81% at Concabella and 93% at Verdú. These values are similar to those obtained by Recasens *et al.* (2001b) (90 and 81%) and by (Recasens *et al.*, 2003) (65 and 92%) in Northeastern Spain, a bit higher to those obtained by Gill and Holmes (1997) (13 and 64%) and a bit lower to those found by Matthews *et al.* (1996) (98%), in Australia. The variation between the experiments suggests that ploughing needs a precise implementation. It should be mentioned that a correct ploughing can be difficult in stony soils like the one in Ferran 2. High fuel consumption justifies its use only in extremely highinfested fields or if cereal price is high enough to compensate the ploughing costs.

Trial results confirm that climatic conditions in the time gap between the normal and the delayed sowing date are crucial for efficacy and should allow maximal *L. rigidum* emergence. Efficacy on weed plants was not considerable at Concabella and Verdú, 35% at Santa Maria, 54% at Ferran 1, and 88% at Sarroca. The highest efficacy was achieved in the field with the highest

Sowing Delay

Sowing delay affected significantly *L. rigidum* density at Ferran 1, Santa Maria and Sarroca. At Ferran 1 and Sarroca, rainfall between both sowing dates allowed new weed emergences, which were mechanically controlled, prior to the second sowing date. At Santa Maria, although little rainfall fell between both sowing dates the soil remained moist due to the abundant rainfall in the months before sowing and to the foggy days, allowing weed emergence. Total weed biomass was also significantly lower in the second sowing date at these three locations. In the other two trials located at Verdú and Concabella, there were no differences in either plant weed density and weed biomass between the two sowing dates. At these two locations too little rainfall fell before the first sowing date and between the two sowing dates to allow *L. rigidum* emergence.

The overall efficacy of sowing delay in these trials was lower than the control obtained by ploughing. The effect of sowing delay on crop density was not statistically significant at Verdú, but at Santa Maria and at Sarroca significantly more crop plants established at the second sowing date ($P < 0.05$). At Santa Maria, no significant differences and even no clear tendencies on *L. rigidum* were detected due to increased sowing rate, so that probably wheat plants were not more competitive when sown in 1.5-fold the normal density. A possible explanation is that the field at Santa Maria was irrigated, consequently, water was not a limiting factor. Increasing sowing density could have a more important effect on weeds in rainfed environments as found by Izquierdo *et al.* (2003) in a barley crop infested with *L. rigidum* who describe a lower weed biomass for the same number of weed plants. Taking into account the yield results of Moreno *et al.* (2002) and Lacasta *et al.* (2004), probably it is not worth to increase the sowing rate more than 160 kg ha-1 in these semi-arid environments.

The Combination of Different Techniques

The theory of using 'many little hammers' to suppress weeds (Liebman and Gallandt, 1997) was partly confirmed at Ferran 1 and also by Recasens *et al.* (2001a). However, the experience of Anderson (2007) in the semiarid steppe of the United States shows that more than two cultural control tactics may be necessary to be effective enough. In the semi-arid conditions of Spain, lack of water in the drylands hinders to integrate other tools proposed by Anderson (2007) as crop rotations of cold-and warm-season crops. In line with the experiences in his situation, no-till could be also used as an effective tool under Spanish conditions. Because seeds remain on the soil surface, they are vulnerable to specific mortality factors that act on surface lying seeds. Seed mortality could contribute declinining the weed seedbank. For example, Baraibar *et al.* (2008) found that predation by harvester ants (*Messor barbarus*) of *L. multiflorum* seeds was significantly higher in no-till than in conventional tilled fields. In case of combining sowing delay with herbicide use, an additional benefit of delayed sowing is that later emerged *L. rigidum* plants are weaker and probably more susceptible to herbicides or to other control strategies (Recasens *et al.*, 2001a).

The starting situation of the fields in the trials described in this work is of highly-infested fields with resistant weeds, which could not be controlled with the conventional herbicide use. In this context, weed density after combining two cultural control methods was still very high ranging between 35 and 450 *L. rigidum* plants m-2 depending on the field. Thus, in highlyinfested fields with herbicide resistant *L. rigidum*, it will probably be necessary to combine these cultural methods with other more drastic tools as leaving the field fallow for one year, depleting the seed bank by repeating cultivation as suggested by Gramshaw and Stern (1977), or combining them with the use of those herbicides that are still active on the populations. As commented previously, the new high cereal prices will probably favour the use of more expensive active ingredients, which should be used correctly in combination with other techniques without relying exclusively on them.

Taking into account that delayed sowing date can have negative consequences on yield due to lack of rainfall and that the effect on *L. rigidum* is more variable, mouldboard ploughing can be a more secure strategy for its control than sowing delay, with more constant effects on weed plant reduction and with a lower impact on yield. The main problem of this technique is the high fuel consumption, which needs to be compensated by the cereal price and by the *L. rigidum* control. Taking into account the fast decline of the seed bank of this species, it is probably useful to plough every 2-3 years, only, to reduce the resistant populations. As the highest reductions were achieved by combining ploughing with sowing delay, it is probably worth to practice both, though high densities of *L. rigidum* will require additional control methods.

Development of Cultural Control Methods for Pests of Leafy Vegetables

This project aimed to evaluate whether or not certain plantings of non-crop plants can assist with pest control in leafy-vegetables such as lettuce and celery. It also assessed the potential of other management practices such as addition of composted fowl manure and using flowering weeds to augment populations of predatory and parasitoids insects and mites. The target pests were primarily aphids and thrips. These are important groups for leafy vegetables as they include lettuce aphid (*N asonovia ribis-nigri*) and western flower thrips (WFT) (*Frankliniella occidentalis*), both of which are resistant to a range of insecticides.

WFT is also an important vector of tomato spotted wilt virus. Strip plantings of cereals will attract certain species of aphids which may be pests in some crops (such as wheat or barley) but are not pests in leafy vegetables. However, predators of cereal aphids such as brown lacewings (*Micromus tasmaniae*) are predators of pest aphids and other insect pests in leafy vegetables.

We aimed to find out whether or not strip plantings of cereals within horticultural crops can (i) increase the populations of beneficial species and (ii) increase the level of control of pests. Mites, staphylinid beetles and predatory thrips are known to prey on thrips and this project investigated how populations of these groups could be encouraged.

It has been noted recently (Beaulieu and Weeks 2007) that there are many soil-dwelling predatory mites that have potential as biological control agents in agriculture, but there has been extremely little research on this group. The use of composted fowl manure was seen as one potential means of achieving increases in resident, soil-dwelling predators. The sequence of events is hypothesised as follows:

1. The application of composted fowl manure will result in more sciarid flies.
2. Predatory mites and beetles that prey on these flies (as eggs and larvae) will be advantaged
3. The predatory mites that are encouraged feed on WFT
4. Therefore, improved control of WFT can be achieved by enhancing generalist predators including predatory mites.

There have been many studies on the provision of pollen and nectar sources to augment populations of biological control agents, particularly hoverflies. There are also some studies where "insectary plantings" have been used to provide an in-field resource to maintain populations of beneficial insects (Colley and Luna 2000). Examples of this type of approach from Australia include the use of strip plantings of lucerne in cotton (Cotton CRC 2005), grasses in citrus (Smith *et al.* 1997) and buckwheat in vineyards (Bernard *et al.* 2006, Berndt *et al.* 2002). In this project we conducted preliminary observations on the potential for existing weeds, particularly flowering aniseed weeds, to provide the same resource for beneficials.

Results

Insectary Plantings and Beneficial Species

The first significant result was that farmers were easily able to understand the concept of insectary plantings and were able to create them using ryegrass. Some farmers planted the entire paddock to cereals and then ploughed them in, others planted all sprinkler rows, and others planted strips throughout the crop. The decision on what sized strip to plant was made by the farmers on the basis of what was practical for them to manage. In all cases we were impressed with the fact that the growers were very happy to plant the cereals once they understood the logic behind the trials. It must be remembered that areas planted to cereals would not produce any crop for them. Sampling of lettuce was undertaken at 4 IPM sites and 6 non-IPM sites. WFT was not abundant in IPM grown lettuce but was found at a much higher level, causing visible feeding damage, on non-IPM lettuce. Two IPM sites were in Werribee South and two were in the Cranbourne area. All of the non-IPM sites were in Werribee South. It ranks the maximum counts of WFT in 10 harvested lettuce during the season from each site as either low (0-25 WFT total), medium (25 – 50 WFT total) or high (50 – 500 WFT).

Discussion and Conclusions

It is clear that there are several useful cultural control measures that are practical, that vegetable growers can use to encourage beneficial insects and mites. That is, the cultural and biological components of an IPM strategy can be strengthened. This project has demonstrated that it is possible to use insectary plantings and made a part of normal vegetable production methods. The use of composted organic material was also shown to increase populations of soil-dwelling predatory insects and mites. The beneficial insects and mites that are encouraged by these different methods fall into two categories, transient and resident. The strip plantings of cereals attract transient brown lacewings (*M. tasmaniae*) as they are dispersing through vegetable production areas in spring. They will stop in these strips but not in a bare paddock, and so it is a means to increase the population of predators on a paddock going into vegetable production. The organic composts can be used to increase the populations of resident soil dwelling predators. The cultural control component of IPM is often given less emphasis than the biological or chemical components but this project has shown that it provides the basis for control of key pests. There are many different types of cultural controls that are available to farmers that are beyond the scope of this project but many are site or crop-specific. This work has focussed on methods to increase populations of beneficial invertebrates. Providing a suitable habitat for predatory species increases the role of both the resident and transient biological control agents. Planting insectary crops has also the added benefit of helping make growers aware of the broader approach to controlling pests by deliberately acting to encourage beneficial species. It is therefore a good starting point for those just adopting IPM, to emphasise the importance of cultural controls within an IPM strategy.

There is also potential for certain existing flowering weeds to be used to increase populations of beneficial insects by providing pollen and nectar. The situation that exists in areas such as Werribee South is that on land adjacent to crops that is usually not owned or managed by the farmer, there are often flowering weeds of different species. Management of these is restricted to limited herbicide applications which can result in bare ground or poor control of established weeds. Replanting these strips of land with native plants may be desirable if particular species can be recommended, but no local species have been trialled in this area and it is currently not achievable by the grower, and is not a priority for them. This especially relevant for farmers leasing land on a short-term basis. We suggest here that there is scope to utilize some existing flowering weeds in preference to attempting to achieve bare earth. The main concern is whether or not certain species of weeds will also encourage pests or be a reservoir of disease. Therefore, in brassica production areas brassica weeds would not be suitable but aniseed weed could be ideal. However, in celery production areas aniseed weed could harbour celery mosaic virus and so would not be suitable. The observations in this project have been that there are often large strips of flowering weeds on land adjacent to crops anyway, but that some of them could be considered useful rather than just weeds.

WFT Control

The studies on WFT in lettuce crops showed that this species was surviving a range of insecticides, including drenches of imidacloprid. It was present in both IPM grown and non-IPM crops but in all sites studied the numbers of WFT never increased in IPM crops. WFT has been present in Victoria for many years but in Werribee South now has the potential to cause serious problems. The small number of IPM lettuce sites available makes firm conclusions difficult, but the results are consistent with the hypothesis that a range of predators, both resident and transient, are present in IPM crops and are preying on WFT, giving control. In the non-IPM crops the number and range of predators is much less and WFT populations have increased. The naturally occurring predators are able to deal with WFT and we suggest that they have been doing so for many years. So there is no requirement to attempt to commercially rear and release these predators outdoors but simply not to kill them where they occur. It is much easier to avoid problems with WFT than to have to deal with this pest when it is in high numbers. Similarly, there is no need to develop specific strategies for WFT. It is dealt with through an effective IPM strategy covering all pests in leafy vegetables.

Chapter 5

Methods to Reduce Nematode Populations

There are numerous general management practices to take into consideration when determining how to manage nematodes. These practices include crop rotation with a nonhost or poor-host crop, resistant varieties, fallowing and chemical controls. Not all of these methods will yield the same results, but they will each provide varying levels of nematode management. Most of these methods, with the exception of chemical controls, will gradually reduce the amount of nematodes in a field, whereas chemical control is more apt to quickly reduce nematodes. When possible, each method should be combined with other methods to produce the best results.

Four Practices Commonly Used to Effectively Combat Nematodes

Crop Rotation	The most important aspect of an effective crop rotation is to find crops that are poor hosts for the nematodes commonly associated with tomatoes. Because numerous species of nematodes are commonly found in one field, it's important to select a crop that is a nonhost for root-knot nematodes, the most damaging nematode in tomato production. Hairy indigo, American jointvetch, sorghum, rye, barley, wheat and pangola digitgrass have shown to be effective in limiting populations of root-knot nematodes.
Resistant Varieties	Resistant varieties are useful for control of some nematodes and diseases in tomatoes. Nematodes fail to develop and reproduce in the roots of resistant varieties, giving tomatoes a better chance to grow as they would normally.

Contd...

	Growers should not rely on the same variety for disease control year after year and should change varieties to slow down the development of new problems. Additionally, not all nematode-resistant varieties of tomatoes have proven to be consistently resistant, so other practices like crop rotation and chemical controls should be used in concert with the resistant varieties.
Fallowing	Having clean land is often a very effective method of reducing nematodes. Populations of nematodes hinge on the availability of a food source; without a crop to feed on, nematode levels will dwindle due to starvation. However, weeds and volunteer crops need to be controlled in a plowed, fallow field as nematodes have a wide host range. If properly managed, a fallow field can decrease the level of nematodes.
Chemical Controls	Nematicides, such as Telone® II soil fumigant, are proven to reduce nematode populations and, in turn, improve fruit production. Telone II is injected into the soil as a liquid and immediately converts to a gas, which creates a zone of protection around developing roots. As a fumigant, Telone II moves through the soil on its own, rather than requiring water or incorporation for movement.

Plant-parasitic nematodes cause major economic losses to horticultural and field crops in the U.S. Current control methods rely largely on nematicides (both fumigant and non-fumigant types) for most crops and pre-plant soil fumigation with the broad-purpose fumigant methyl bromide for high value fruit and vegetable crops. However, there are serious concerns about the use of nematicides in terms of food safety and environmental quality issues.

In addition, methyl bromide is proposed to be withdrawn from use in the U.S. in 2005. Alternative environmentally compatible control strategies that use bio-based options (including resistant, non-host, or antagonistic crop cultivars and biological control agents) are urgently needed to replace high-risk chemical nematicides.

Importance of work and consequences if not done. Identification, characterization, and enhancement of resistant crop germplasm, antagonistic crops cover crops, and biocontrol agents are essential for the development of bio-based management options against plant-parasitic nematodes.

Assessment of the effects of such management techniques on diversity of both plant-parasitic and free-living nematodes will increase our understanding of soil health and aid in development of sustainable crop management.

If this work is not done, crop production will continue to move offshore where environmental standards for pesticide use are less stringent and our supply of safe, inexpensive grains, fruits, and vegetables will dwindle. Moreover, local farmers will lose market share and have reduced opportunities for economic development.

Nematode Management in Organic Agriculture

Nematodes are usually microscopic in size and are classified as unsegmented worms, belonging to the Phylum Nematoda. Plant-parasitic nematodes are a concern for growers of agricultural or garden crops. These plant-parasitic nematodes will mainly feed on the roots of plants. A few kinds will feed on foliage but this not common. Many other kinds of nematodes are present in the soil as well.

These include decomposers, predators, insect parasites, and animal parasites. Some nematodes are aquatic and do not affect terrestrial plants. Other nematodes act as decomposers, predators, and insect parasites. In farming systems, nematode predators and parasites of insects are beneficial, while nematode parasites of animals and plants are considered pests in agriculture.

Beneficial nematodes that feed on either bacteria or fungi help in nutrient cycling by accelerating the decomposition of organic matter. Predatory nematodes may keep harmful nematodes at lower levels by feeding on them. Entomopathogenic nematodes (nematodes harmful to insects) may help to reduce numbers of some insect pests by infecting them with bacteria.

The objectives of this publication are to provide information on plant parasitic nematodes causing damage in organic agriculture and to introduce methods for their management.

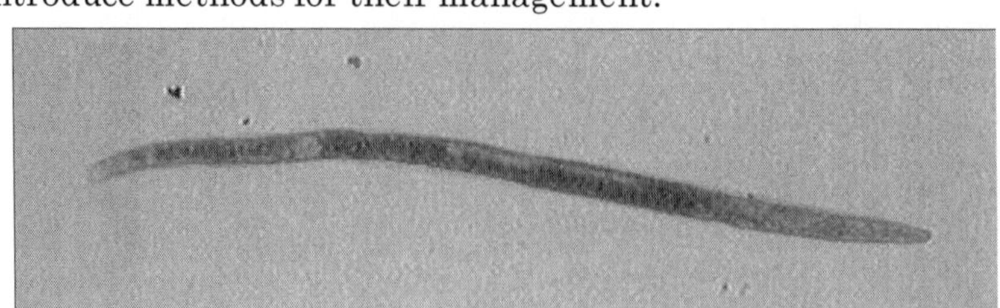

Figure: *A typical plant parasitic nematode.*

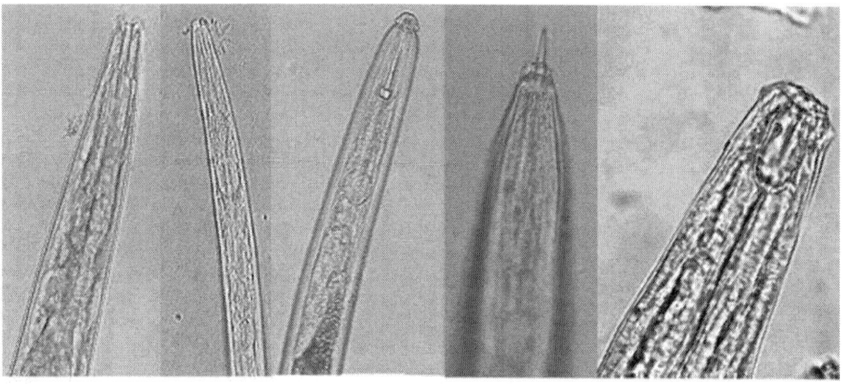

Figure: *Different nematodes: a) Bacterivore, b) Fungivore, c) Herbivore, d) Omnivore, e) Predator*

Nematode Damage and Detection

Since most plant-parasitic-nematodes feed on plant roots, symptoms are comparable to nutrient or water deficiency. These can be yield loss, stunting, yellowing, wilting, symptoms of nutrient deficiency, and malformations of the root (including tubers, peanuts) caused by direct feeding damage. In addition, invasion by plant-parasitic nematodes often provides an infection route for other organisms such as bacteria or fungi, since nematode activity creates an entryway into the root that would otherwise not be available.

Nematode reproduction is very fast. On average, a typical life cycle may be only 30 days at summer temperatures. This means that even if nematode numbers are low at the beginning of the growing season, nematode populations can rapidly increase and can become harmful to the crop in a relatively short period of time.

Often, the most damaging nematodes in the southeastern United States and the tropics are root-knot nematodes (*Meloidogyne* spp.). These nematodes are pests of nearly all major crops and are therefore widespread. Damage can be directly observed by examining the roots, because root-knot nematodes produce galls or knot-like swellings along the plant roots.

These galls cannot not be easily removed because they are part of the plant root tissue. In contrast, nodules caused by beneficial nitrogen fixing bacteria can be easily removed. Another method to distinguish nematode galls from nodules is to slice them in half and expose to air, which will cause nitrogen-fixing nodules to appear pink or red if the bacterial colony is active, or green or brown if the colony is inactive.

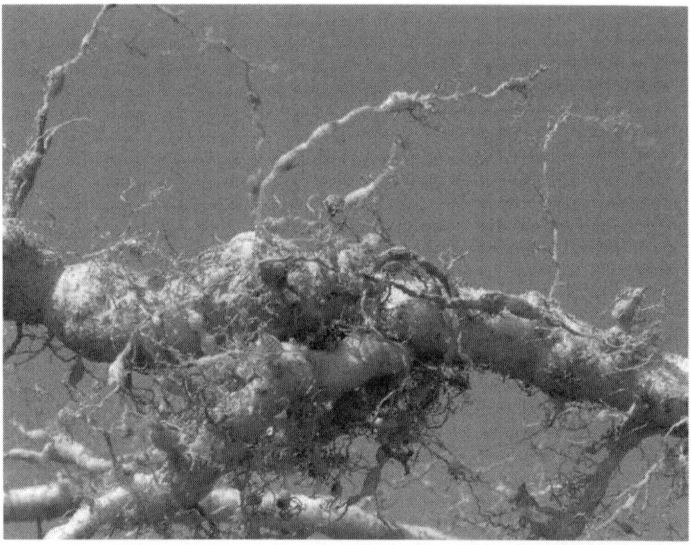

Figure: *Root galls caused by root-knot nematodes.*

Evidence of direct feeding, apparent in the form of galls and other root malformations leads to the damage symptoms described above. The invasive stage of the root-knot nematode life cycle is the juvenile root-knot nematode, which can freely move through the soil and will enter the root of a suitable host plant. Only the female is capable of establishing a feeding site. The female will become immobile, causing the plant to form giant cells for feeding, which essentially appears as a gall on the root. The presence of root-knot nematodes can be detected by the presence of these characteristic galls.

Soil Samples. Other plant-parasitic nematodes do not cause these gall symptoms, and so soil samples should be taken to determine which nematodes are present in a field. Even when it is clear that root-knot nematodes are to blame for the damage caused, it may be beneficial to find out if other nematodes are present to determine which treatments should be employed and will work best.

Nematode Management

In nematode management it is important to remember that nematodes can move only very short distances on their own. Therefore nematodes are mainly spread through lack of sanitation and movement of infected soil and planting material. In order to limit a build-up of nematodes, planting equipment and tools should be properly cleaned, and in extreme cases could only be used for the same field. Furthermore only soil and planting material free of nematodes should be used, because once nematodes are introduced into a field they cannot be eradicated. After harvest infected plants should be destroyed to prevent the build-up of nematodes on these crop residues and therefore in the soil. Once they become established in a site, the nematodes will persist there, and management will be required on a regular basis.

Resistant Plants and Rotation Crops. Nematode management is primarily a pre-planting activity. In order to protect the crop most activities must be started two or three months before the scheduled planting date. Several pre-plant treatments are available for the organic farmer. The choice of a suitable crop cultivar can be a critical decision. Host plant resistance achieved by traditional breeding programmes can be a valuable protection against some nematodes.

Two terms that are often used when talking about host plant resistance to nematodes are "tolerance" and "resistance." Tolerance means that the plant can withstand some damage caused by nematodes without experiencing significant yield reduction. In contrast, resistance means that nematode reproduction is very low or non-existant on the plant. Both provide protection for the crop plant, but the next crop following a tolerant plant could be damaged by the nematodes that survived on the tolerant plant. Different plant species or even cultivars of the same plant species can exhibit varying degrees of resistance or tolerance.

According to the University of Missouri-Columbia extension service, the following vegetable plants are recommended as reasonably resistant to root-knot nematodes: broccoli, brussel sprouts, mustard, chives, cress, garlic, leek, groundcherry, and rutabaga. In contrast, Globe artichokes, Jerusalem artichoke, asparagus, sweet corn, horseradish, some lima bean varieties, onion, and rhubarb are considered to be tolerant.

However, the Alabama Cooperative Extension System provides additional information on resistant crops. Cowpea or black-eyed peas, which can be used as a rotation crop to reduce nematode numbers, has several cultivars that are quite different in their ability to support reproduction of the root-knot nematode Meloidogyne incognita. Planted into the sandy soils of Florida, the cultivars Tennessee Brown, Mississippi Silver, and California Blackeye #5 inhibited reproduction more effectively than the cultivar Purple Knuckle (intermediate). Population densities were high after planting the cultivars Whippoorwill, Pinkeye Purplehull, and Texas Purplehull. Iron Clay is another cowpea cultivar that is known to reduce population levels of M. incognita.

Crop rotation utilizes crops that are a poor or non-host to the nematodes found in an agricultural field. These crops can either be plants that provide a secondary cash crop grown in between cycles of the primary cash crop, or they could be cover crops that are not primary but provide benefits to the farming system such as nitrogen enrichment, nematode reduction, or possible additional income. In either case, nematode numbers are reduced simply because nematodes are deprived of a suitable host crop.

This does not mean that nematode densities are reduced indefinitely, but a successful crop rotation should reduce nematode levels enough so that a following susceptible crop will produce sufficient yields and survive until the end of its regular growing season. Popular cover crops are sorghum, sorghum- sudangrass, different grains such as oat and rye, many grasses, marigold, cowpea, and some tropical legumes such as sunn hemp (*Crotalaria juncea*) and velvetbean (*Mucuna* spp.). These cover crops are useful to reduce root-knot nematode population densities. Even though these cover crops are effective against root-knot nematodes, they might increase other harmful plant-parasitic nematodes that are also present in the soil.

For example, densities of the sting nematode (*Belonolaimus longicaudatus*) and stubby-root nematode (*Paratrichodorus minor*) increased on SX-17 sorghum-sudangrass. Mississippi Silver cowpea was able to reduce *P. minor* and *M. incognita* but increased *B. longicaudatus*. The most effective cover crop that simultaneously reduced root-knot, stubby-root, and sting nematodes was determined to be velvetbean. Results vary in different locations as well. In Kenya, a plant related to sunn hemp (*C. ochroleuca*), which successfully reduced populations of root-knot nematodes appeared to increase the occurrence of lesion nematodes (*Pratylenchus zeae*), whereas jack bean (*Canavalia ensiformis*), hyacinth bean (*Lablab purpureus*), and velvet bean appeared to reduce both nematodes.

Response of nematodes is greatly influenced by the cultivar of a crop plant species. Therefore even if the same plant species is used, results may vary if the cultivar is different. Sorghum is often recommended as a cover crop to decrease population levels of root-knot nematodes, and is widely used for this purpose.

However, use of an uncommonly used cultivar (Asgrow Chaparral) actually increased population levels of root-knot nematodes (*M. incognita*) in a field experiment in North Florida. Furthermore, reactions of plants seem also to be site dependent. Cover crops that decrease nematode populations in one field may not be effective in another, depending on local nematode species and their local populations and variants present. Since weeds can harbor nematodes, it is important to have a dense cover crop stand that blocks out weeds.

Table: *Cover crops, which demonstrated low or intermediate susceptibility to* M. arenaria *race 1,* M. incognita *race 1, and* M. javanica .

Common Name	***Scientific Name***	***Cultivar***
cowpea	*V. unguiculata*	'Mississippi Silver' 'Iron Clay'
sunnhemp	*Crotalaria juncea*	
American jointvetch	*Aeschynomene americana*	
marigold	*Tagetes minuta*	
	T. patula	'Dwarf Primrose'
velvetbean	*Mucuna deeringina*	
sorghum-sudangrass	*Sorghum bicolour x S. sudanese*	'SX-17'
jack bean	*Canavalia ensiformis*	
partridge pea	*Cassia fasciculata*	
sesame	*Sesamum indicum*	'Sesaco 16' 'Paloma'

Tillage

Tillage and the practice of fallowing fields may appear as alternatives to cover crops for nematode management. Tillage inverts and mixes soil and exposes deeper soil layers to the sun. This practice is meant to kill nematodes by desiccation, since nematodes depend on moisture for survival. This practice may kill some of the nematodes that are in the upper soil layers, however it will not reach nematodes that have retreated into moderate or deeper soil layers. Nematodes can retreat to depths greater than 12 inches (30 cm) in Florida soils, and can migrate upward once a susceptible host is planted.

Once a field has been fallowed, nematodes will move into deeper soil layers to avoid drying and may enter an inactive stage that enables them to survive periods without food and in addition protects them from desiccation. McSorley and Gallaher (1993) compared the effects of tillage versus crop rotation on nematode densities in tropical corn (*Zea mays*) cultivar Pioneer X304C in North Florida. Tillage did not have significant effects on nematode densities, while a rotation crop of sorghum cultivar DeKalb BR64 reduced population levels of root-knot nematodes (*M. incognita*). These results demonstrate that tillage is not a reliable method for nematode management in Florida. Fallowing can deprive nematodes of food, if the field is kept 100 % weed free, and therefore can reduce population levels. But clean fallowing is a poor practice for soil conservation and nutrient runoff. In the absence of plants that keep soil in place, wind will be able to carry some of the topsoil away. Additional soil will be lost with rainwater runoff. Rain will also carry soil nutrients into lakes and rivers, which will degrade water quality and interfere with aquatic ecosystems.

Solarization. A promising technique is the use of heat to decrease not only nematode densities, but also other harmful organisms and weed seeds. This can involve pasteurization, steaming, or solarization of the soil before planting. Of these, solarization is probably the most practical. It involves the covering of the soil with clear plastic. Transparent plastic sheets allow short-wave radiation from the sun to penetrate the plastic. Once the light passes through the plastic and is reflected from the soil, the wavelength becomes longer and cannot escape through the plastic.

The trapped light facilitates heating of the soil to temperatures detrimental to most living organisms. There are different types of plastic sheets available, mainly differing in their thickness (insulation) and ability to let light through (transparency). Black, opaque, or translucent plastics are not suitable for solarization. Thin, transparent plastic sheets appear to achieve the best results. Katan (1981) recommended thicknesses between 25 to 30 μm.

Temperatures can rise anywhere between 35 and 60 C (95-140 F) during the summer months when air temperatures are close to 32 C (89.6 F) or higher. However, cloud cover and rain limit solar radiation and may diminish the success of solarization. Soil temperatures only rise to detrimental levels in the first 10 to 30 cm of soil (Katan 1987), and even in this range temperatures drop off as depth increases. The plastic has to be sealed to prevent air movement underneath the plastic, which would prevent temperatures from rising sufficiently.

Solarization

The soil should remain covered for a minimum of four weeks, but increasing solarization time improves effectiveness. This helps to heat the soil at a greater depth, which means that more nematodes will be affected by it. Furthermore it also ensures that an adequate accumulation of solarized hours is achieved, which is important in regions like Florida where the sky is often overcast during the summer months.

A combination of a suitable cover crop and solarization seems to achieve best results. Wang et al. (2006) showed that following a cowpea cover crop with solarization accomplished better results than solarization alone. In fact, the effectiveness of this combined treatment was comparable to methyl bromide fumigation, which in conventional agriculture was the most effective treatment used against soil-borne diseases and pests.

The disadvantage of solarization is its negative impact on beneficial soil organisms, since they will meet the same fate as their harmful counterparts. But recovery is usually attained quickly through rapid recolonization. Furthermore, other beneficials such as *Bacillus, Pseudomonas*, and *Trichoderma* are able to survive the high temperatures generated by solarization. Solarization generates considerable plastic waste, which, when farming with a sustainable approach in mind, is a drawback. However, plastic sheets could be used multiple times.

Biological Control. Biological control is the management of plant-parasitic nematodes by living organisms such as bacteria, fungi, predatory nematodes, or other invertebrates. Biological control is mainly accomplished by attempting to build-up beneficial organisms through the use of various soil amendments. The introduction of beneficial soil organisms to the soil has only been attempted successfully in a few instances. *Pasteuria* spp., which are bacterial parasites of various plant-parasitic nematodes occurring naturally in Florida soils, may be promising biological control agents. In a 7-year experiment with tobacco, Weibelzahl-Fulton et al. (1996) showed that *M. incognita* and *M. javanica* were suppressed exclusively by *P. penetrans.*

Tobacco plants treated with *P. penetrans* had fewer galls, egg masses, and eggs than plants that either did not receive *P. penetrans* or lacked the biological control agent due to autoclaving the soil in which the plants were grown. Nematode-trapping fungi are also potential candidates for biological control. Their adhesive knobs, rings, or net structures trap nematodes and kill them. Other types of fungi may parasitize nematode eggs.

Neither fungi nor *Pasteuria* spp. are available for widespread commercial use at this moment. But they can be found in a healthy soil environment, as can predators such as mites and predatory nematodes. Recent research has shown that a wide range of nematode natural enemies occur in Florida soils. Most predators are generalists, meaning they feed on a variety of prey species during their lifetime. For biological control purposes, generalist predators are a disadvantage, because they will not only feed on the key pest, but on other suitable organisms as well including beneficials. Furthermore if there is not enough food available they will disperse, possibly causing the pest to rebound. On the other hand, generalist predators can keep many different kinds of pests at low population densities. Currently, no predators are commercially available for augmentive releases for nematode control in vegetable production systems.

Organic Amendments in Relation to Biological Control

Biological control is difficult in soil, because it is a complex environment. Many of the possible organisms that could provide biological control lack specificity and therefore will not focus on a particular organism and may even interfere with beneficials. Therefore biological control of nematodes is achieved mainly by conservation of existing biological control; meaning that the soil environment is modified to aid the survival and reproduction of nematode natural enemies that are already present. Primarily this is accomplished through the addition of organic amendments. Organic amendments can improve the soil environment to aid biological control, benefit general plant health by helping with water retention and providing additional nutrients, and affect nematodes directly and negatively through detrimental decomposition products.

The impact of organic amendments on nematodes is often inconsistent and unpredictable. In most cases when organic amendments are applied, they are helpful mainly as a plant nutrient source and do not directly aid in nematode management. However, even if nematodes are unaffected by the added amendments, plant health may improve due to other favourable properties of amendments. McSorley and Gallaher (1995) observed that amendments consisting of yard waste compost (four to six-month-old wood fragments, leaves, grass clippings) used as mulch or incorporated into the soil improved yield of squash and okra even though plants experienced heavy galling. Although nematode densities in composted treatments were similar to those in non-amended soil plants benefited greatly from the improved retention of soil moisture. Kimpinski et al. (2003) showed similar results. Amendments (compost and manure) increased yield in barley and potatoes in Canada, but this was not necessarily due to reduced nematode numbers.

The addition of organic amendments may stimulate the entire soil food web, including beneficial biological control organisms. When soil is amended with nutrient-rich organic matter, microorganisms may immobilize the released nutrients. Free-living nematodes that feed on these bacteria and fungi may play an important role in the release of nutrients tied up in such a way. Soil-inhabitating microorganisms and nematodes therefore may act together to influence the rate of decomposition and release of nitrogen from crop residues, including nitrogen-rich leguminous crops like sunn hemp. Wang et al. (2004) investigated effects of sunn hemp decomposition on the soil nematode community. Litterbags that allowed nematodes to pass in and out were filled with sunn hemp hay and buried in soil. Periodic analysis of the bags showed that sunn hemp decomposition was fairly rapid and most was completed within two weeks.

Population densities of nematodes feeding on bacteria increased greatly during decomposition, and this was followed by an increase in omnivorous nematodes that fed on them. However, plant-parasitic nematodes were neither affected by the sunn hemp decomposition nor by the omnivorous nematodes.

In fact, plant-parasitic nematode numbers increased over time with increasing crop biomass, which may indicate that sunn hemp residue does not have a direct detrimental effect on plant-parasitic nematodes. In another recent study, Chellemi (2006) used urban plant debris (green waste from public landfill deposited by homeowners and landscape companies) to amend pepper fields in southeast Florida. As a result, the combined density of plant-parasitic nematodes was reduced, and other disease-causing plant pathogens (*Pythium* spp. and *Phytophthora* spp.) were reduced as well. These examples illustrate the potential use of organic amendments for direct or indirect management of nematodes, but they also indicate that outcomes can be complex and unpredictable.

Management of Infected Plants

Once plants are infected with nematodes, there is little that can be done to remove or reduce nematodes. Therefore, prevention and sanitation are critical to controlling nematodes. The improvement of plant health is an important cultural technique to lessen detrimental effects on plants caused by plant-parasitic nematodes. Proper irrigation, fertilization and organic amendments as surface mulches or soil incorporated are important. Furthermore, removal of weed hosts and old crop plants immediately after harvest can reduce nematode densities for the future. Nematodes are a long-term pest, which cannot be eradicated once they become established in a site. They can only be kept at low levels with carefully selected management tactics that are often specific to the managed site.

There are a variety of additional methods that may have some effect on nematodes. These include methods such as use of rhizobacteria, chitin, sesame residues, flooding, or microwave energy. Some of these (flooding, microwaving) may be restricted to specialized situations. Performance of these methods may be inconsistent or ineffective in some situations. Amendments such as chitin or sesame residue may provide nutrients that are beneficial to overall plant health regardless of any effects on nematodes. In these cases only small test areas should be subjected to new materials to determine their effectiveness under local conditions. Many of the methods described above such as host plant resistance and rotation are more dependable under a variety of conditions.

Nematode Management in Tomatoes, Peppers, and Eggplant

Plant parasitic nematodes, are small microscopic roundworms that live in the soil and attack the roots of plants. Crop production problems induced by nematodes, therefore, generally occur as a result of root dysfunction, reducing rooting volume and foraging and utilization efficiency of water and nutrients. Many different genera and species of nematodes can be important to crop production in Florida. In many cases a mixed community of plant parasitic nematodes is present in a field, rather than having a single species occurring alone. In general, the most widespread and economically important nematode species include the root-knot nematode, *Meloidogyne* spp., and sting nematode, *Belonolaimus longicaudatus*.

Figure: *The simultaneous occurrence of both root-knot nematode (*Meloidogyne *spp.) and Fusarium wilt disease (*Fusarium oxysporum*) causing enhanced disease development and tomato yield loss.*

The host range of these nematodes, as with others, includes most if not all of the commercially grown vegetables within the state. Yield reductions can be extensive but vary significantly between plant and nematode species.

In addition to the direct crop damage caused by nematodes, many of these species have also been shown to predispose plants to infection by fungal or bacterial pathogens or to transmit virus diseases, which contribute to additional yield reductions.

Biology and Life History

Most species of plant parasitic nematodes have a relatively simple life cycle consisting of the egg, four larval stages, and the adult male and female. Development of the first stage larvae occurs within the egg where the first molt occurs.

Second stage larvae hatch from eggs to find and infect plant roots or in some cases foliar tissues. Host finding or movement in soil occurs within surface films of water surrounding soil particles and root surfaces. Depending on species, feeding will occur along the root surface or in other species like root-knot, young larval stages will invade root tissue, establishing permanent feeding sites within the root. Second stage larvae will then molt 3 times to become adult male or female.

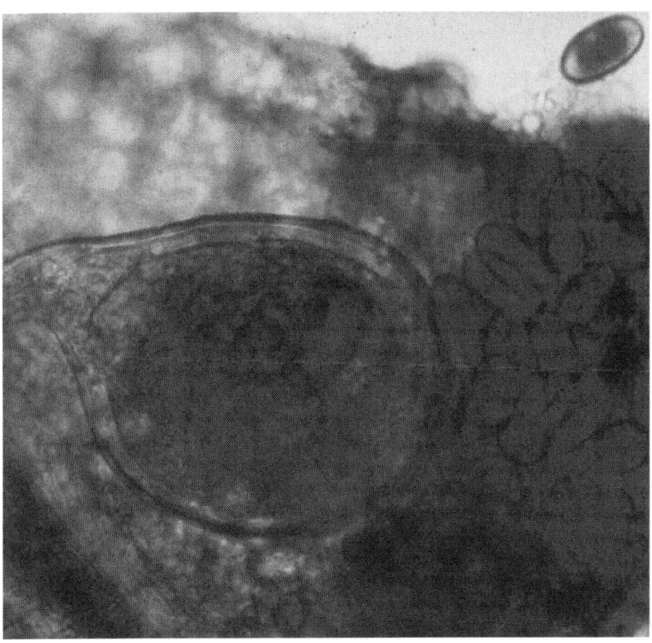

Figure: *Adult female of the root-knot nematode (*Meloidogyne *sp.) with attached egg mass.*

For most species of nematodes, as many as 50-100 eggs are produced per female, while in others such as root-knot, upwards of 2000 may be produced. Under suitable environmental conditions, the eggs hatch and new larvae emerge to complete the life cycle within 4 to 8 weeks depending on temperature. Nematode development is generally most rapid within an optimal soil temperature range of 70 to 80°F.

Symptoms

Typical symptoms of nematode injury can involve both aboveground and below ground plant parts. Foliar symptoms of nematode infestation of roots generally involve stunting and general unthriftiness, premature wilting and slow recovery to improved soil moisture conditions, leaf chlorosis (yellowing), and other symptoms characteristic of nutrient deficiency.

An increased rate of ethylene production, thought to be largely responsible for symptom expression in tomato, has been shown to be closely associated with root-knot nematode root infection and gall formation. Plants exhibiting stunted or decline symptoms usually occur in patches of nonuniform growth rather than as a overall decline of plants within an entire field.

Figure: *Wilting, yellowing, and stunting of pepper due to the root-knot nematode (*Meloidogyne *sp.). Note the formation of galls on roots.*

The time in which symptoms of plant injury occur is related to nematode population density, crop susceptibility, and prevailing environmental conditions.

Figure: *Wilting of pepper due to massive root infection by the root-knot nematode (*Meloidogyne *sp.).*

For example, under heavy nematode infestation, crop seedlings or transplants may fail to develop, maintaining a stunted condition, or die, causing poor or patchy stand development.

Under less severe infestation levels, symptom expression may be delayed until later in the crop season after a number of nematode reproductive cycles have been completed on the crop.

In this case aboveground symptoms will not always be readily apparent early within crop development, but with time and reduction in root system size and function, symptoms become more pronounced and diagnostic.

Root symptoms induced by sting or root-knot nematodes can oftentimes be as specific as aboveground symptoms. Sting nematode can be very injurious, causing infected plants to form a tight mat of short roots, oftentimes assuming a swollen appearance.

New root initials generally are killed by heavy infestations of the sting nematode, a symptom reminiscent of fertilizer salt burn. Root symptoms induced by root-knot cause swollen areas (galls) on the roots of infected plants.

Gall size may range from a few spherical swellings to extensive areas of elongated, convoluted, tumorous swellings which result from exposure to multiple and repeated infections. Symptoms of root galling can in most cases provide positive diagnostic confirmation of nematode presence, infection severity, and potential for crop damage.

Damage

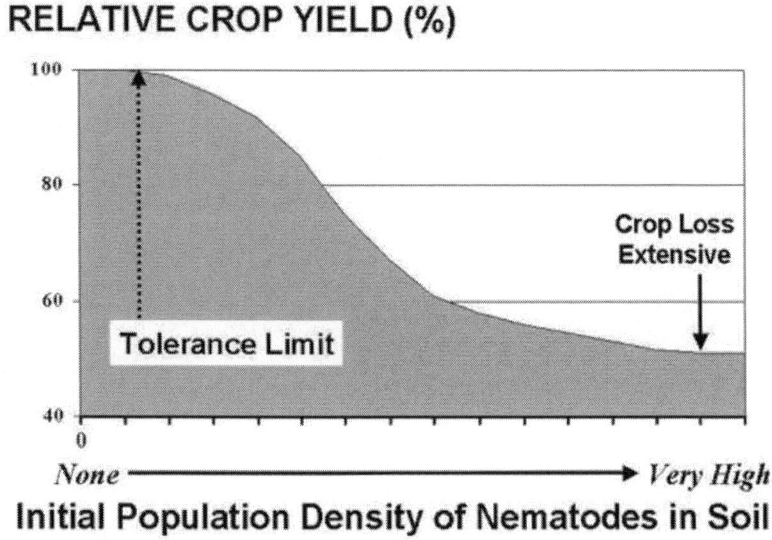

Figure: *Typical nematode induced crop damage relationship in which crop yields, expressed as a percentage of yields that would be obtained in the absence of nematodes, decline with increased population density of nematodes in soil. The tolerance level is identified as the initial or minimal soil population density at which crop damage is first observed.*

For most crop and nematode combinations the damage caused by nematodes has not been accurately determined. Most vegetable crops produced in Florida are susceptible to nematode injury, particular by root and sting nematodes. Plant symptoms and yield reductions are often directly related to preplant infestation levels in soil and to other environmental stresses imposed upon the plant during crop growth. As infestation levels increase so then do the amount of damage and yield loss. In general, the mere presence of root-knot or sting nematodes suggests a potentially serious problem, particularly on sandy ground during the fall when soil temperatures favour high levels of nematode activity. At very high levels, typical of those that might occur under doubling cropping, plants may be killed. Older transplants, unlike direct seed, may tolerate higher initial population levels without incurring as significant a yield loss.

Field Diagnosis and Sampling

Because of their microscopic size and irregular field distribution, soil and root tissue samples are usually required to determine whether nematodes are causing poor crop growth or to determine the need for nematode management. For nematodes, sampling and management are a preplant or postharvest consideration because if a problem develops in a newly planted crop, currently no postplant corrective measures are available to rectify the problem completely once the nematode becomes established.

Nematode density and distribution within a field must therefore be accurately determined before planting, guaranteeing that a representative sample is collected from the field. Nematode species identification is currently only of practical value when rotation schemes or resistant varieties are available for nematode management. This information must then be coupled with some estimate of the expected damage to formulate an appropriate nematode control strategy.

Advisory or Predictive Sample (Prior to Planting)

Samples taken to predict the risk of nematode injury to a newly planted crop must be taken well in advance of planting to allow for sample analysis and treatment periods if so required. For best results, sample for nematodes at the end of the growing season, before crop destruction, when nematodes are most numerous and easiest to detect. Collect soil and root samples from 10 to 20 field locations using a cylindrical sampling tube or, if unavailable, a trowel or shovel.

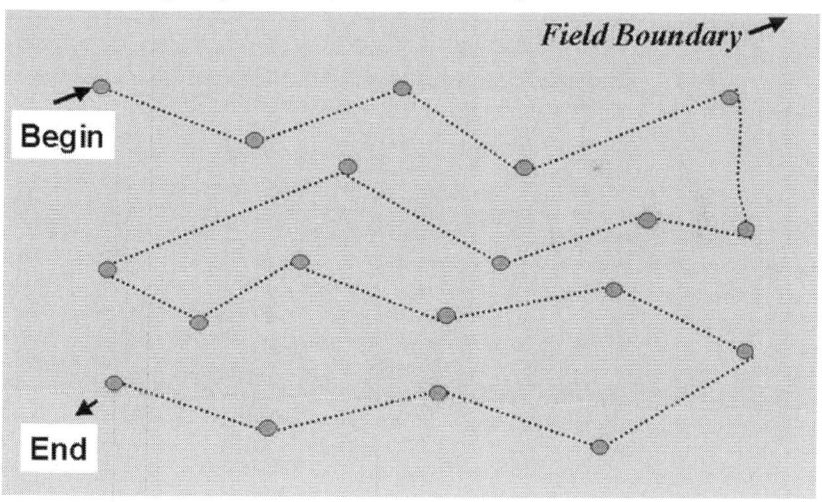

Figure: *Figure: Suggested pattern for collecting preplant soil samples for nematode analysis based on compositing soil from 10-20 field locations.*

Since most species of nematodes are concentrated in the crop rooting zone, samples should be collected to a soil depth of 6 to 10 inches. Sample in a regular pattern over the area, emphasizing removal of samples across rows rather than along rows. One sample should represent no more than 10 acres for relatively low-value crops and no more than 5 acres for high value crops. Fields which have different crops (or varieties) during the past season or which have obvious differences either in soil type or previous history of cropping problems should be sampled separately. Sample only when soil moisture is appropriate for working the field, avoiding extremely dry or wet soil conditions.

Diagnostics on Established Plants

Roots and soil cores should be removed to a depth of 6 to 10 inches from 10 to 20 suspect plants. Avoid dead or dying plants, since dead or decomposing roots will often harbor few nematodes. For seedlings or young transplants, excavation of individual plants may be required to insure sufficient quantities of infested roots and soil. Submission of additional samples from adjacent areas of good growth should also be considered for comparative purposes.

For either type of sample, once all soil cores or samples are collected, the entire sample should then be mixed thoroughly but carefully, and a 1 to 2 pint subsample removed to an appropriately labelled plastic bag. Remember to include sufficient feeder roots. The plastic bag will prevent drying of the sample and guarantee an intact sample upon arrival at the laboratory. Never subject the sample(s) to overheating, freezing, drying, or to prolonged periods of direct sunlight. Samples should always be submitted immediately to a commercial laboratory. If sample submission is delayed, then temporary refrigerated storage at temperatures of 40 to 60°F is recommended.

Figure: *Suggested strategy for collecting post-plant soil samples for nematode analysis comparing sampling results from areas of good and poor plant growth.*

Recognizing that the root-knot nematode causes the formation of large swollen areas or galls on the root systems of susceptible crops, relative population levels and field distribution of this nematode can be largely determined by simple examination of the crop root system for root gall severity. Root gall severity is a simple measure of the proportion of the root system which is galled. Immediately after final harvest, a sufficient number of plants should be carefully removed from soil and examined to characterize the nature and extent of the problem within the field. In general, soil population levels increase with root gall severity.

Figure: *Field diagnosis for the presence and severity of root-knot nematodes via visual examination of uprooted plant root systems for root galls.*

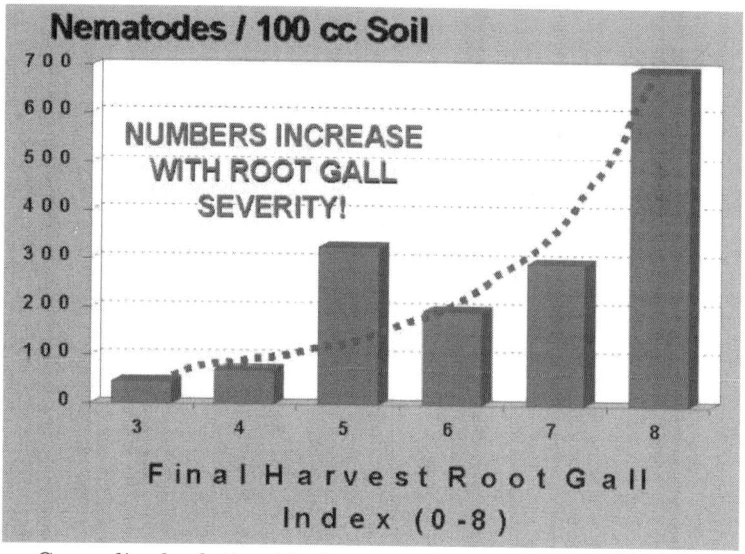

Figure: *Generalized relationship between the number of root-knot nematode juveniles in soil based on the level and severity of plant root galling.*

This form of sampling can in many cases provide immediate confirmation of a nematode problem and allows mapping of current field infestation. As inferred previously, the detection of any level of root galling usually suggests a nematode problem for subsequent planting of a susceptible crop, particularly within the immediate areas from which the galled plant(s) were recovered.

General Management Considerations

Currently nematode management considerations include crop rotation of less susceptible crops or resistant varieties, cultural and tillage practices, use of transplants, and preplant nematicide treatments. Where practical, these practices are generally integrated into the summer or winter 'off-season' cropping sequence. It should be recognized that not all land management and cultural control practices are equally effective in controlling plant parasitic nematodes and varying degrees of nematode control should be expected.

These methods, unlike other chemical methods, tend to reduce nematode populations gradually through time. Farm specific conditions, such as soil type, temperature, and moisture, can be very important in determining whether different cultural practices can be effectively utilized for nematode management.

Cultural Practices

Crop Rotation

For crop rotation to be effective, crops unsuitable for nematode infection, growth, or reproduction must be introduced into the rotation sequence. In most of Florida it is not uncommon to observe a multispecies community of nematodes all occurring within the same field. Under these circumstances it may not be possible to find a rotation or cover crop that will effectively reduce populations of all nematode pests, particularly if root-knot and sting nematodes occur in combination In this case, crop rotations detrimental to root-knot, which is generally the most difficult to control, should be selected. In some cases, resistant crop varieties are available, which can be used within the rotation sequence to minimize problems to some species of root-knot but not sting nematodes.

Use of poor or nonhost cover crops within the rotation sequence, may in some cases offer an effective approach to nematode control. Four leguminous cover crops adaptable for managing soil populations of sting or root-knot nematode include Iron Clay Cowpea (*Vigna unguiculata* . cv. 'Iron Clay'), sunn hemp (*Crotalaria juncea),* hairy indigo (*Indigofena hirsuta*) and American jointvetch (*Aeschynomene americana*). Sorghum is also a popular cover crop restoring large amounts of soil organic matter, but is a good host for sting nematode but not root-knot. Most of the small grains commonly used as winter cover crops in central and north Florida, such as rye, barley, wheat, or oats, can support limited reproduction of root-knot nematode. To avoid an increase in root-knot populations, these crops should only be planted when soil temperatures are below 65°F, a threshold temperature for nematode activity. Cover crop rotations with some pastureland grasses (particularly pangola digitgrass, and to some extent bahiagrass, and Bermuda grass) have significantly reduced, but not eliminated, root-knot nematodes. In north Florida, long term (6- to 9-year) pastureland rotations have allowed economic melon production within root-knot infested fields. It should be recognized that as the crop rotation period is shortened or eliminated, nematode problems will intensify accordingly. Other perennial legumes currently under evaluation may play an important role in future nematode management programmes.

For cover crops to be most effective stands must be established quickly and undesirable weeds that can serve as alternative hosts must be controlled. Given that many different weeds serve as alternative plant hosts to nematodes (i.e., nutsedges), it may not be possible to manage root-knot nematode with crop rotation unless an integrated programme to manage weeds is also considered and implemented within the field. With many cover crops, rapid stand establishment has been a significant problem. Similarly, economic crop rotation sequences are often further complicated by lack of crop management skills, specialized equipment to grow and harvest the crop, or by the lack of closely located processing facilities or markets. In some cases other measures should be considered such as fallowing, which is usually as efficient as crop rotation for reducing field infestations of nematodes.

Fallowing

Clean fallow during the off-season is probably the single most important and effective cultural control measure available for nematodes. When food sources are no longer readily available, soil population densities of nematodes gradually decline with death occurring as a result of starvation. Due to the wide host range of many nematode species, weeds and crop volunteers must be controlled during the fallow period to prevent nematode reproduction and further population increase. At least two discing operations are generally required to maintain clean fallow soil conditions during the interim period between crops. Fallowing by use of herbicides to deplete nematode populations is a much slower process because the soil is not disturbed, thereby subjecting nematodes from deeper soil layers to the drying action of sun and wind. The unfavourable effects of fallowing on soil organic matter and soil structure is usually more than compensated for by the level of nematode control achieved and the resulting increase in crop productivity. When soil erosion is a potentially serious problem other measures should be considered.

Biological Control

At present there are no effective, commercially available, biological control agents which can be successfully used to control nematodes.

Plant Resistance

Use of nematode-resistant crop varieties has not been extensively evaluated in Florida, but is often viewed as the foundation of a successful integrated nematode management programme on all high value crops in which methyl bromide is currently used. Commercially available nematode-resistant varieties are currently available only for tomato, pepper, southernpea, and sweet potato. In a resistant variety, nematodes fail to develop and reproduce normally within root tissues, allowing plants to grow and produce fruit even though nematode infection of roots occurs. Some crop yield loss can still occur, however, even though the plants are damaged less and are significantly more tolerant of root-knot infection than that of a susceptible variety.

In tomato, a single dominant gene (subsequently referred to as the Mi gene) has been widely used in plant breeding efforts and varietal development, which confers resistance to all of the economically importance species of root-knot nematode found in Florida, including *Meloidogyne incognita, arenaria,* and *javanica.* Commercially resistant fresh market varieties, climatically and horticulturally adapted for Florida, are available in a number of tomato varieties (e.g, 'Sanibel'). Unfortunately, in previous research with resistance tomato varieties, the resistance has often failed as a result of the heat instability or apparent temperature sensitivity of the resistant Mi gene. For example, previous research has demonstrated threshold soil temperatures and incremental reductions in nematode resistance with each degree above 78°F, such that at 91°F tomato plants are fully susceptible. This would suggest that in Florida use of these varieties may have to be restricted to spring plantings when cooler soil temperatures prevail.

In pepper, two newly developed root-knot nematode resistant varieties (Carolina Belle and Carolina Wonder) were released from the USDA Vegetable Research Laboratory for commercial seed increase in April 1997. Both varieties are open pollinated, and homozygous for the N root-knot nematode resistant gene. Preliminary research has demonstrated that these varieties confer a high degree of resistance to the root-knot nematode; however, expression of resistance is heat sensitive. Further research is necessary to characterize the usefulness of these varieties under the high soil temperature conditions of Florida. Like tomato, use of these varieties may have to be restricted to spring plantings when cooler soil temperatures prevail.

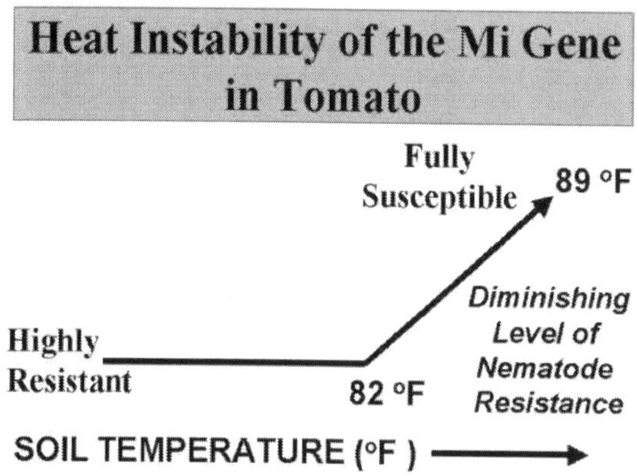

Figure: *Diagrammatic representation showing the complete loss of root-knot nematode resistance conferred by the Mi gene in tomato with increasing soil temperature.*

In addition to problems of heat instability, the continuous or repeated planting of resistant plant varieties will almost certainly select for virulent races of *Meloidogyne* capable of overcoming the resistance. Therefore, the duration and/or utility of the resistance may be time-limited. In previous studies with resistant tomatoes, resistance breaking nematode races have been shown to develop within 1 to 3 years. Since new races of the nematode can develop so rapidly, a system of integrated control usually mandates the rotation of resistant and non-resistant varieties to slow the selection process for new virulent races. Previous trials in Florida have demonstrated the capacity of some species or races of root-knot to reproduce and inflict damage upon a resistant tomato variety.

The results of these experiments have demonstrated that even with a resistant variety, which were damaged less than that of a susceptible variety, some consideration of initial soil population levels of the root-knot nematode must be observed to minimize tomato yield losses. Given that significant yield losses can still occur, combined efforts to manage soil populations to low levels prior to planting must still be considered, particularly if tomatoes are planted as a fall crop.

If this situation develops, the combination of a nematicide and resistant variety may also comprise an option to reduce nematode populations to nondamaging levels.

Soil Amendments

Many different types of amendments and composted materials have been applied to soil to suppress populations of plant parasitic nematode and improve crop yield and plant health. Animal manures, poultry litter, and disk-incorporated cover crop residues are typical examples of soil amendments used in agriculture to improve soil quality and as a means for enhancing biocontrol potential of soil. Some amendments that contain chitin and inorganic fertilizers that release ammoniacal nitrogen into soil suppress nematode populations directly and enhance the selective growth of microbial antagonists of nematodes. More recently, composted municipal wastes and sludges have been used to amend soil to improve soil fertility, organic matter content, water-holding capacity, nutrient retention, and cation exchange capacity.

Suppression of soilborne pathogens via the incorporation or simple mulching of composted amendments is reputedly based on enhanced microbial activity and increased numbers of antagonists generated by decomposition of the amendment in soil. Soils with a diversity of beneficial microorganisms are more suppressive to pathogens than soils with little or no biological diversity.

Other possible mechanisms for pathogen suppression by composts include direct inhibition of the pathogen or reduced infectivity of the organisms into the plant host. Population increases of beneficial organisms in soil appear to be the direct result of environmental changes brought about by the amendments after addition to soil. This suggests that to sustain soil suppressiveness, amendments must be periodically reapplied to maintain the soil environment conducive to antagonists.

The level to which soilborne pest and disease control can be achieved is not only related to the type of material but to the age of the compost. Nematode and disease suppression has been repeatedly demonstrated with composted municipal yard wastes containing significant quantities of tree bark.

If the compost is immature, the product may not only be difficult to handle and have an offensive odor, but may contain salts and metabolites toxic to plants. For example, weed suppression has been demonstrated with some types of immature composted materials that contain and or produce organic acids with phytotoxic properties. Other studies have shown that soils amended with different sources of composted municipal wastes were disease suppressive as the long as they were relatively fresh (< 6 months), but as the composted municipal waste was aged, disease suppressiveness was lost.

In other Florida studies, application of composted municipal wastes at rates up to 120 tons per acre have not been shown to be pesticidal in activity, but actually dramatically increased populations of nematodes and other disease organisms such as *Fusarium* and *Phytopthora* spp. Nematode population increases were directly related to increases in plant growth and root system size with amendment application rate.

Previous studies in Florida were conducted to determine the extent to which increasing application rates of a municipal solid composted waste affect the ability of tomato plants to tolerate root infection by species of root-knot nematode (*Meloidogyne* spp.) These studies showed that in a sandy soil poor in organic matter content (less than 2%), tomato yields could be increased significantly with soil amendments in both nematode free or nematode infested soil.

The impact of the root-knot nematode on tomato yield was effectively constant, however, suggesting that application of the soil amendment did not enhance the ability of tomato plants to tolerate infection by the root-knot nematode. Much of the previous and ongoing research in Florida also seems to indicate that the major effects of soil amendments to crop yields appear to be less related to nematode or soil pathogen control than to enhanced plant nutrition and nutrient and water availability.

It is not clear at this time and preliminary stage of university field research whether benefits to crop growth after the initial crop following soil amendment application can be expected. Previous studies showed no response in second crop tomato yields (double crop) following amendment application rates from 15 to 120 tons per acre. Disappearance of nutrients and soil organic matter content appears to be very rapid in the hot, moist soils of Florida. Preliminary research suggests that reapplication of the amendments may have to be made on 1-2 year basis to sustain crop growth and yield benefits. In summary, the high rates of application (tons/acre) and attendant costs required for crop response and nematode control for many different types of organic amendments, and the apparently rapid losses of the materials in soil appear to preclude use of these materials primarily to homeowner or small farm operations at this time. However, with additional research and advances in application technology and use efficiency, use of soil amendments may become an integral component of Florida crop production systems.

Flooding

Flooding has been shown to suppress nematode populations. Alternating 2 to 3 week cycles of flooding and drying have proven to be more effective than long, continuous flooding cycles. At present, only limited areas within the state are situated to take advantage of flooding as a viable means of nematode control. Given the growing concern about aquifer depletion, salt water intrusion, and water use inefficiencies, it seems unlikely that Florida water management officials will continue to permit flooding within these areas in the future.

Soil Solarization

Soil solarization is a nonchemical technique in which transparent polyethylene tarps are laid over moist soil for a 6 to 12 week period to heat noncropped soils to temperatures lethal to nematodes and other soil-borne pathogens. Soil temperatures are magnified by the trapping of incoming solar radiation under the clear, polyethylene panels. To be effective, soils must be wetted and maintained at high soil moisture content to increase the susceptibility (thermal sensitivity) of soil borne pests and thermal conductivity of soil. Wet mulched soils increase soil temperatures due primarily to the elimination of heat loss by evaporation and upward heat convection, in addition to a greenhouse effect by prohibiting dissipation of radiation from the soil. At the end of the solarization period the clear plastic is painted with a white latex paint to allow continued use of the plastic as a mulch cover for the production of vegetables on raised beds.

Soil solarization, the use of clear plastic panels as bed coverings to heat non-cropped soils to temperatures lethal to nematodes and other soilborne pathogens.

Soil Decontamination

A 2008 study used a solar cell to generate an electric field for electrokinetic (EK) remediation of cadmium-contaminated soil. The solar cell could drive the electromigration of cadmium in contaminated soil, and the removal efficiency that was achieved by the solar cell was comparable with that achieved by conventional power supply. In Korea, various remediation methods of soil slurry and groundwater contaminated with benzene at a polluted gas station site were evaluated, including a solar-driven, photocatalyzed reactor system along with various advanced oxidation processes (AOP). The most synergistic remediation method incorporated a solar light process with TiO2 slurry and H2O2 system, achieving 98% benzene degradation, a substantial increase in the removal of benzene.

Attempts were made to use solar energy for controlling disease agents in soil and in plant material already in the ancient civilization of India. In 1939, Groashevoy, who used the term "solar energy for sand disinfection," controlled Thielaviopsis basicola upon heating the sand by exposure to direct sunlight.

Soil solarization is the third approach for soil disinfestation; the two other main approaches, soil steaming and fumigation; were developed at the end of the 19th century. The idea of solarization was based on observations by extension workers and farmers in the hot Jordan Valley, who noticed the intensive heating of the polyethylene-mulched soil. The involvement of biological control mechanisms in pathogen control and the possible implications were indicated in the first publication, noticing the very long effect of the treatment. In 1977, American scientists from the University of California at Davis reported the control of Verticillium in a cotton field, based on studies started in 1976, thus denoting, for the first time, the possible wide applicability of this method.

The use of polyethylene for soil solarization differs in principle from its traditional agricultural use. With solarization, soil is mulched during the hottest months (rather than the coldest, as in conventional plasticulture which is aimed at protecting the crop) in order to increase the maximal temperatures in an attempt to achieve lethal heat levels.

In the first 10 years following the influential 1976 publication, soil solarization was investigated in at least 24 countries and has been now been applied in more than 50, mostly in the hot regions, although there were some important exceptions. Studies have demonstrated effectiveness of solarization with various crops, including vegetables, field crops, ornamentals and fruit trees, against many pathogens, weeds and a soil arthropod. Those pathogens and weeds which are not controlled by solarization were also detected. The biological, chemical and physical changes that take in solarized soil during and after the solarization have been investigated, as well as the interaction of solarization with other methods of control. Long-term effects including biological control and increased growth response were verified in various climatic regions and soils, demonstrating the general applicability of solarization. Computerized simulation models have been developed to guide researchers and growers whether the ambient conditions of their locality are suitable for solarization.

Studies of the improvement of solarization by integrating it with other methods or by solarizing in closed glasshouses, or studies concerning commercial application by developing mulching machines were also carried out. The use of solarization in existing orchards (e.g. controlling Verticillium in pistachio plantations) is an important deviation from the standard preplanting method and was reported as early as 1979.

How to Solarize Soil

The area to be solarized should be level and free of debris and large clods. The soil should be watered if the soil is dry. The soil should not be saturated but have the crumbly damp look that the soil would have at planting. Place clear plastic over the soil surface and bury the edges in a foot deep trench. The plastic can be clear construction grade plastic and vary in thickness from 1 to 6 mils. The thinner plastic (1-2 mils) will allow better soil heating since it will reflect less solar energy. The plastic should be left in place for 4-6 weeks. Solarization is most effective when done in June and July. However, depending on your geographic location, solarization may also be effective in May, August, and September.

The most successful use of soil solarization appears to occur in heavier (loamy to clay soils) rather than sandy soils. Soils with poor water holding capacity and rapid drainage can significantly inhibit heat transfer to deeper soil horizons. Loss of pest control is directly correlated with soil depth. The depth to which lethal temperature can be achieved (6 to 8 inches) is also dependent on the intensity and duration of sunlight and ambient temperature. At present, the only time to consider soil solarization for pest control is during our hot, summer and early fall months, which fortunately are 'off-season' in most peninsular Florida vegetable row crops. Unfortunately our summers are also our wettest period of the year with frequent afternoon rain showers that have a cooling effect on the soil.

Many different pests have been suppressed and or controlled by soil solarization, particularly within arid environments with intense sunshine, and limited cloud cover and rainfall. Previous studies in Florida have demonstrated that soil solarization can also be effective in a subtropical environment. Plant parasitic nematodes have generally proved to be more difficult to control with soil solarization, as have some weed pests such as crabgrass in a central Florida study. The results of preliminary experiments also suggest the potential for selection pressures towards a buildup of heat tolerant individuals that may serve to reduce soil solarization efficacy after repeated use as a nematode control tactic. In some studies, effective use of solarization for nematode control has required an integrated systems approach, coupling solarization with other chemical or nonchemical approaches. For example, the combined use of soil solarization with a nematicide has improved nematode control and crop yield.

In addition, use of virtually impermeable, photo-selective plastic mulches may also complement low dose fumigant treatments to reduce weed germination and growth in the event of extended periods of cloud cover occurring during the solarization regime. At this time, further research is needed demonstrating soil solarization pest control activity and consistency in the various geographical regions of Florida where vegetable crops are grown.

Other Cultural Practices

Other cultural measures that reduce nematode problems include rapid destruction of the infested crop root system following harvest. Fields that are disced as soon as possible after the crop is harvested will not only prevent further nematode population growth but subject existing populations to dissipation by sun and wind.

Use of nematode free tomato transplants is also recommended since direct seeded plants are particularly susceptible since they are vulnerable to injury for a longer duration, during an early, but critical period of crop development. Since nematodes can be carried in irrigation water that has drained from an infested field, growers should avoid use of ditch or pond waters for irrigation or spray mixtures.

In most cases, a combination of these management practices will substantially reduce nematode population levels but will rarely bring them below economically damaging levels. This is especially true of lands that are continuously planted to susceptible crop varieties. In these cases some form of pesticide assistance will still usually be necessary to improve crop production.

Antagonistic Crops

A large proportion of the food production in many tropical and subtropical countries comes from smallholders farming relatively small parcels of land. Crops may be grown together in mixed cultivations often following a system that has been long established and generally successful. In the era of "progress" signalled by the development of monoculture with new varieties, crop protection aids and inorganic fertilizers, it is regrettable that in hindsight the knowledge of yesterday contained in the practices and methods of these smallholders has not been given more attention.

The trend towards urbanization has generated the need for regular supplies of food to be produced from farms and smallholdings close to cities, or with reasonable access to a road network. Intensification of food-crop production to meet this consumer demand has caused various problems for agronomists, soil scientists and crop protectionists, particularly if the land is in continuous cultivation.

The traditional system of agriculture relied on the use of long periods of fallow between crops. During this time the natural vegetation regenerated, soil fertility was restored and populations of pests and diseases declined. Traditional farming systems are usually of mixed crops (even including trees or woody perennials) that provide an ecosystem in which total crop loss due to pests and diseases is rare. Nevertheless, subsistence farmers will usually grow enough to cover the possibility of losing up to 40 percent of production before or after harvest. The problems of a rapidly increasing human population for land use has led to the abandonment of traditional systems, and these are now practised only in sparsely populated countries.

Crop losses due to pests and diseases appear to be more frequent as improved crop varieties are grown in monoculture on the same land with the aid of pesticides and inorganic fertilizers. In many instances, crop damage has increased as a result of the inadvertent introduction of alien pests on crop plants. The distribution of several nematode species has been increased in this way, e.g. *Radopholus similis* and *Pratylenchus* spp. on bananas, *Tylenchulus semipenetrans* on citrus and *Globodera* spp. on potatoes.

In considering potential strategies for controlling nematodes there are certain practices that can be implemented once the life history and the behaviour of the pest is known. Plant-parasitic nematodes occur in greatest abundance in the rooting zones of crops, mostly in the top 30 cm of soil; nematodes may also penetrate to deeper soil layers particularly in unfavourable conditions (Prot, 1980) only to return to the surface layers when crop growth resumes. The free-living stages of some species survive the unfavourable season through physiological or structural adaptation, or survive for several seasons as eggs, as in the case of the potato-cyst nematode.

Nematodes are usually at their most vulnerable at the time when they are actively searching for host roots and when surviving unfavourable growing seasons. The survival of free-living stages in the absence of hosts can be longer than originally thought, and fallow periods of less than 12 months may be insufficient to control those species that reproduce rapidly. The main abiotic factors that influence free-living nematodes and eggs are heat, desiccation and anaerobis. Plant-parasitic nematodes will not survive long periods at temperatures above 40°C. These temperatures may be reached on bare soil in some countries during the hot season, but heat penetration may not extend below 10 cm.

The technique of increasing soil temperatures by solarization can be of value in controlling nematodes, but this is only possible where there are long periods of uninterrupted sunshine. Solarization is also practised for controlling soil-borne fungal pathogens. Nematodes are aquatic animals that inhabit the films of water that surround soil particles. As soils dry, the volume of soil water decreases and the ability of nematodes to move is impaired, eventually free-living nematodes may die as the soil moisture decreases.

The extent to which species have adaptations to survive desiccation is still a matter for further study. By regulating the metabolism in times of stress some species are capable of anhydrobiotic survival. The contrivance of bare fallows that deprive nematodes of a host may not always be a worthwhile control strategy especially as fallowing in hot, dry areas can deplete soil organic matter and the levels of beneficial soil biota. In such situations for specific nematode pests, a better alternative might be cover crops with non-host leguminous trees or shrubs such as *Crotalaria* spp.

The control of nematodes by flooding has been advocated in certain locations. When flooded all the soil pore spaces are filled and the oxygen supply for the soil microflora and fauna becomes limiting. Many plant-parasitic nematodes are intolerant of oxygen starvation and soon die. Similar effects are observed when nematodes are stored in deep water in the laboratory, but often they can be revived by aeration. To be effective in the field therefore, the anaerobiosis has to be of sufficient duration to kill the nematodes. In Côte d'Ivoire banana fields are flooded for at least five weeks. Flooding is not a technique that can be widely used for nematode control and whether it is equally efficacious against all plant-parasitic species is not known.

The development of strategies for nematode control requires a thorough understanding of the growth of the host plant and the biology of the nematode. In some climates it is possible to avoid (or contain) damage by planting the crop at a time when nematode invasion will be slow, or when plant growth will be faster than that of the parasite. Damage to transplanted short-cycle crops can be limited if seedlings are produced in nematode-free nurseries or in soil-less potting media. In general, protection from invasion during the early growth of a crop will enable the plant to produce a yield acceptable T most farmers even if it is planted in nematode-infested soil, particularly if the crop receives adequate supplies of water and nutrients.

This will not prevent the development of large nematode populations on the root systems and in the soil that will influence the subsequent crop. Brown showed that treating wheat with low nematicide doses at seeding gave a substantial yield improvement, resulting from protection from the invasion of cereal-cyst nematodes during the establishment of the crop.

The concept of growing companion plants with nematicidal root exudates alongside susceptible crops is attractive, but the actual benefits may be questionable or at least difficult to quantify. Species of *Tagetes, Crotalaria* and castor bean (*Ricinus communis*) have been widely used. Where *Tagetes* was grown as a companion to tomato in a rotation trial in the United Republic of Tanzania, yields were significantly less than with other treatments, indicating that such a cropping practice was not likely to be adopted by farmers.

The use of organic mulches for managing nematodes has been widely studied and there are various examples of plant debris having beneficial effects on plant growth. In small garden plots that receive heavy mulches of organic waste, nematodes are unlikely to be a major problem. Subsistence gardens are therefore well suited to the needs of householders, who, as Hussey (1990) reported, will usually grow sufficient quantities to meet immediate needs and allow for crop losses due to pests and diseases. Regular mulching is realistic for subsistence gardeners but becomes less feasible if the area of cultivation is large. To be effective, the recommended mulches may have to be used at application rates of 5 to 10 tonnes per hectare. The acquisition and distribution of such volumes is not likely to be practised by farmers growing crops on a large scale.

Future Developments

If chemical treatments are unavailable or inappropriate, the management of plant-parasitic nematodes will have to depend on alternative strategies. Those farmers in the Near East region who routinely use methyl bromide or non-fumigant nematicides may in future have to seek alternative treatments. To help them do so, government-supported research will be necessary, particularly for control strategies not requiring the regular purchase of proprietary products. This type of research is unlikely to be funded by manufacturing companies.

New research will require multidisciplinary collaboration to enable the integration of traditional practices with new ideas created by the scientists.

Through more research on nematode biology and the environment, nematologists must consider several possibilities:

- How nematode life cycles may be interrupted.
- How microbial activity at the root-soil interface, which could lessen nematode invasion, can be promoted.
- Identification of compounds that interfere with reception of stimuli or repel nematodes.
- The selection or breeding of resistant varieties, or varieties that tolerate nematode infestation, and that can produce acceptable yields.

In conducting field experiments that evaluate the effect of cropping systems on nematode populations, scientists should plan to suit the needs of the farmer. Many research station trials are designed to produce results for ready analysis and presentation in journals; in consequence, treatments may be selected to suit the needs of the scientist. Most farmers grow short-cycle, annual and perennial crops in close proximity. Such mixed cropping practices are rarely undertaken in field experimentation by nematologists.

Any new cultural, microbial or management approaches that are developed must be within the capabilities of the farmer and meet the necessary environmental and economic requirements. New strategies will be considered as successful if they lower nematode damage thresholds. The farmer will be satisfied if these treatments are reliable, practicable and economically justified; his customer, the consumer, will be content if the product has the desired quality, contains no toxic residues, and is fairly priced.

Biofumigation

Bio-fumigation is based on incorporating soil amendment (fresh plant mass, manure) into the soil, which will release chemical substances, known as isothiocyanates (ITC's), able to suppress soil-borne pests and diseases, plus a soil heater to enhance biological activities. Plants from Cruciferae family (cabbage, radish, cauliflower etc.) release large amount of these toxic to soil-borne pests and diseases substances - in the soil and are considered the best material for bio-fumigation.

How to do it:

- Incorporate the fresh mass into the soil. This can be done directly if the mass is coming from grown crop or plant mass taken from elsewhere and brought into the plot or field. If the mass is transported to the field, the soil should be well prepared before the incorporation. During transportation and storage of these organic materials in the field, care must be taken not to lose the gases produced from biodegradation, by covering the piles of the bio-fumigant with plastic until the time of application.

A dose of 50 t / ha is recommended, although when problems with nematodes or fungi are very serious, 100 t / ha should be applied, a dose that can be reduced by choosing a cultivation techniques such as application in furrows.

The bio-fumigant should be distributed uniformly, so that no concentration of pathogens will appear that could create problems for the crop. Once the biofumigant is distributed, it should be incorporated immediately into the soil.

- Water the field, if possible by sprinkling, until the soil is saturated, although watering can be done by flooding, or drip irrigation can be installed.
- Cover the soil surface tightly with a transparent plastic film for at least 2 weeks to retain the gases produced from the biodegradation of the organic matter. This could be the same plastic as the one used for soil solarisation.
- The film is removed 3-4 weeks after and the soil slightly removed in order to permit the gases to escape from soil.
- Planting of the interested crop can be done 24 hours later.

Using Bio-fumigation against Nematodes

Different mustards (e.g. Brassica juncea var integrifolia or Brassica juncea var juncea) should be used as intercrop on infested fields. As soon as mustards are flowering they are mulched and incorporated into the soil. While incorporated plant parts are decomposing in a moist soil, nematicidal compounds of this decomposing process do kill nematodes. Two weeks after incorporating plant material into the soil a new crop can be planted or sown (it takes about 2 weeks for the plant material to decompose and stop releasing phytotoxic substances = chemicals poisonous to plants).

Mexican marigold, also known as Tagetes, has been successfully used in the control of root-knot nematode in roses by a Kenyan Farmer.

It is recommended to alternate the use of agricultural residues with green manure, especially from brassicae, using 5-8 kg square-metres of green matter, although combinations of legumes and grass can be applied. In the case of the use of green manure cultivated in the same field, fast growing plants should be used to be incorporated at least 30 days after having been planted, to avoid the increase of pathogen populations.

Planting mustard after bio-fumigation can serve as bio-indicators of possible phytotoxicity, because the germination of their seeds is sensitive to phytotoxic substances. At the same time they are very sensitive to nematodes and permit the detection of areas in the crop where biofumigation is not effective. They act like trap plants, and like bio-fumigants when incorporated into the soil.

In Spain, successful application of bio-fumigation was achieved in strawberries, peppers, cucurbits, tomato, brassicae, cut flowers, citrus and banana. Bio-fumigation has also been recently applied to Swiss chard and carrot crops. The most utilized biofumigants have been goat, sheep and cow manure, and residues from rice, mushroom, brassicae and gardens.

The effectiveness of bio-fumigation in controlling nematodes, fungi, insects, bacteria, and weeds is nearly the same as with the use of conventional pesticides. Bio-fumigation may also regulate viral problems by controlling vector organisms.

Biofumigation Against Bacterial Wilt

From 1999 research on using bio-fumigation for control of bacterial wilt has been carried out in Australia and the Philippines by the Autralian Centre for International Agricultural Research (ACIAR). The Centre with collaborators have been working to identify suitable brassicas for tropical environments and to evaluate them in the field, initially on experimental farms and more recently on commercial farms and smallholder farmers fields. There are many varieties and they differ enourmously in the level of disease-suppressing chemicals they produce. In recent years Australian banana growers have started growing brassica green manures for nematode control, and in the US farmers are using mustard green manure crops to replace synthetic fumigation for potatoes with huge savings on costs.

Growing of radish, mustard and broccoli have reduced bacterial wilt significantly (50-60%) in most of the experiments, though researchers believe the treatment is more effective on sandy soil than on heavy clay soil.

Methods to Increase Microbial Abundance Andor Activity

Nematophagous fungi parasitize nematodes. The possibility of using these fascinating fungi for biological control of plant- and animal-parasitic nematodes has resulted in increased interest in their biology. The following images can be used to introduce this concept to students.

Endoparasitic fungi use their spores to infect nematodes. The spores adhere to the nematode cuticle or, in some species, are ingested together with food. The endoparasitic fungi are obligate parasites of nematodes and spend their entire vegetative lives inside infected nematodes. Examples of these fungi are *Catenaria anguillulae* (zoospores), *Drechmeria coniospora* (adhesive conidia), and *Nematoctonus concurrens* (adhesive spores).

Nematode-trapping fungi use hyphal trapping devices to capture nematodes. These traps can have either adhesive or mechanical function. In contrast to the endoparasites, the nematode-trapping fungi can live as saprophytes in soil. Examples of nematode-trapping fungi with different trapping devices are *Arthrobotrys oligospora* (adhesive networks), *Dactylaria candida* (adhesive knobs), *Monacrosporium cionopagum* (adhesive branches), *Arthrobotrys dactyloides* (constricting rings), and *Nematoctonus leiosporus* (hourglass traps).

Chapter 6

Public Health Pests and Disease Vectors

Public health pests are organisms which attack or annoy us in some way. When pests such as insects attack us by biting or stinging, disease symptoms may result. Some animal species harbour pathogenic microbial organisms such as bacteria or viruses. If an animal transmits such an organism to a person, and the person becomes ill, then the animal has served as a disease vector. Some organisms are included as public health pests not because they directly attack or annoy us, but because they help produce situations favouring the growth of pests or vectors. Weeds are an example of this public nuisance. California Health and Safety Code (2002 (j)) defines public nuisances as:

(1) Any property, excluding water, that has been artificially altered from its natural condition so that it now supports the development, attraction, or harborage of vectors. The presence of vectors in their developmental stages on a property is prima facie evidence that the property is a public nuisance.

(2) Any water that is a breeding place for vectors. The presence of vectors in their developmental stages in the water is prima facie evidence that the water is a public nuisance.

(3) Any activity that supports the development, attraction, or harborage of vectors, or that facilitates the introduction or spread of vectors.

A century ago it was easy to make distinctions between arthropods that were strictly pests and those which were vectors of disease organisms. Before 1920 only a few infectious diseases were known to be associated with arthropods. In fact, it had been only since 1876, when Dr. Patrick Manson demonstrated that mosquitoes were involved in the life cycle of filarial worms causing filariasis in humans that any insect appeared on a list of vectors.

In the years since then, vector biologists and epidemiologists have increased this list many-fold, and further have shown that the ecology of many vector-pathogen relationships is very complex, and that vectors can have a number of roles in pathogen transmission and maintenance. Many arthropod species formerly considered only pests are now known to play vital roles in maintenance, amplification, and transmission of vertebrate pathogens. Thus, with the realization that certain arthropods can cause diseases other than infectious diseases in humans and other animals comes the conclusion that there are really few significant distinctions to be made between arthropod pests and arthropods of public health importance.

A core function in solving a problem caused by public health pests and disease vectors is proper identification of the suspected pest species. Often, this is a difficult task best left up to highly specialized biologists called taxonomists. Identifications made to the level of species can be of vital importance to understanding the habits, developmental sites, and susceptibility of the pests to control measures, but sometimes knowing basic biological characteristics of the major group to which the pest belongs can be extremely helpful. It is only after gaining experience working with groups of pests that pesticide applicators can know which pests can be identified by sight or with field guides, and which pests will need to be sent to specialists for identification. Failure to properly identify a pest causing a problem may result in failure to solve the problem, and thus wasted time, money, chemicals, and effort. Proper identification will make it possible for you to understand much more about the pest, to make appropriate selection and application of management techniques in a way that solves the problem with minimal secondary effects to people and the environment.

Taxonomists divide living organisms into two large groups: the Animal Kingdom and the Plant Kingdom. A third Kingdom containing single-celled organisms such as bacteria and viruses are recognized by most taxonomists, but it is difficult to find agreement for the name of this Kingdom and just what should be included in it. One name that many taxonomists use for this Kingdom is the Protista. Most public health pests and disease vectors can be placed into two groups in the Animal Kingdom, the invertebrates (animals without backbones) and vertebrates (animals with backbones) and the loosely-defined category called weeds in the Plant Kingdom. This is the arrangement that will be followed in this manual.

How to Identify Pests

Although details will vary according to the type of pest to be identified, the basic principles of identification are the same for all pest species.

1. Compare collected specimens of the pest with identified specimens in museums or with photographs or other illustrations of identified pests.
2. Use an illustrated key. Keys are published guides to identification of biological organisms where the reader is led through a series of choices (usually two choices) on a certain characteristic of a specimen. If each choice is made correctly, the reader is eventually led to the correct identification of the specimen. Keys can be difficult to use for non-specialists because the characteristics contained in the choices are not always obvious. Keys that use choices such as "legs black" or "legs yellow" are easy to use. Those that ask "subcostal vein bifurcate" or "subcostal vein not bifurcate" are not.
3. Contact an authority, such as a farm advisor, agricultural commissioner, or biologist in the USDA, Cooperative Extension, public health department, or at a college or university.

Invertebrates

Invertebrate animals include millions of animals that are not pests, such as sponges, starfish, earthworms, and squid. Nearly all the public health invertebrate pests are in the Phylum Arthropoda, and are called arthropods. The arthropods of public health importance are primarily the insects and their close arachnid relatives ticks, mites, and spiders.

Insects and Related Arthropods

Insects

There are well over one million species of insects, which means that there are more kinds of insects than any other kind of organism. Of these million-plus species, fewer than 100,000 are found in the United States. Of these, only a small number are important as public health pests. In California we have about 100 common insect species of public health importance. Insects may cause diseases or annoyance to people in many ways, including:

- Causing allergic reactions
- Causing fright and annoyance
- Crawling, scraping, biting, bloodsucking, and other ways of attacking people
- Invading body tissues or organs of people
- Stinging people
- Transmitting disease organisms
- Serving as hosts to disease organisms.

Development of Insects

Insects have complex life cycles. Each insect changes form one or more times during its life. This change of form, called metamorphosis, may be divided into two types—simple and complete. Insects with simple metamorphosis hatch from eggs and appear to be miniature copies of the adult insect: these are called nymphs. As the nymphs feed and grow, they shed their skin several times — a process called molting. When they complete their development, they are adults. Many common insects have simple metamorphosis, including cockroaches, termites, lice, thrips, true bugs, and aphids.

The more complex type of development among insects is called complete metamorphosis. This developmental pattern involves several complete changes in form during the growth of the insect. Insects with complete metamorphosis begin life as an egg. A larva hatches from this egg; the larva feeds, grows and molts several times. Then the larva enters a stage called a pupa. In this stage the insect assumes form unlike that of the larva or the adult. Pupae usually show little movement, but mosquitoes have an aquatic pupa that moves to the water surface to breath and sinks to bottom to hide. Mosquito pupae are called "tumblers".

In all insects with complete metamorphosis the adult insect develops within the pupa and eventually an adult insect emerges. It is the adult that mates, produces eggs, and starts the next generation. Common examples of insects with complete metamorphosis are beetles, moths and butterflies, flies, fleas, bees and ants. Insects having complete metamorphosis present a special challenge to identification because larval stages usually look little like their adult forms.

Insect Form and Structure

Adult insects have a segmented external skeleton, three major body regions, six jointed legs, one pair of antennae, complex mouth parts, and (frequently) two pair of wings. The "skin" of an insect is in fact the segmented external skeleton, the exoskeleton, which covers the entire body.

This exoskeleton protects the insect from enemies, gives the animal rigidity, and serves as the support of the soft internal body parts, including the muscles, nerves, blood system and various organs and glands. The exoskeleton is jointed to allow the animal to move, but it cannot grow. Therefore the exoskeleton must be shed from time to time to permit growth and development.

Adult insects have three major body regions: a head, a thorax, and an abdomen. The insect head bears the sensory organs — eyes, antennae, and mouthparts. The eyes of insects may be simple cells capable only of light detection, or they may be highly complex organs made of thousands of individual elements, capable of detecting movement, shapes and colours.

The thorax of the insect bears most of the organs of locomotion, including legs and wings. The legs may be highly modified for such activities as running, digging, or swimming. The wings, if present, may be simple and underdeveloped, or they may be highly developed for rapid and powerful flight.

The abdomen of the insect may have varying forms, but in the adult it lacks legs or wings. It may have some sensory organs on the hind end, but these are never true legs. The abdomen bears the reproductive organs of both the male and female insects.

Adult insects usually have two pair of wings, although certain groups, such as the true flies, have only one pair while fleas lack wings altogether. Insect wings may be almost featureless and transparent, like those of many flies, or they may be covered with brightly coloured scales, like butterflies. In beetles, wings may be hardened into tough convex body coverings called elytra. The number of wings and their form and ornamentation are important factors in insect identification.

Insect mouth parts are as variable as the wings. The mouthparts may be modified. for chewing plants, shaped into siphon-like tubes for sucking plant juices, developed into needle-like structures for piercing the skin of plants or animals, or modified into sponge-like structures for lapping up liquids.

Taxonomists place insects in the Class Insecta, and further divide them into about 30 orders. Mites, ticks, and spiders are arthropods related to the insects. Generally these insect relatives can be recognized by their overall appearance and by their eight-jointed legs.

Their bodies are divided into two major regions rather than three, and they never have wings.

Arthropods of Public Health Importance: Insects

Figure: *Order Anoplura, a sucking louse*

- No wings
- Piercing-sucking mouthparts
- Simple metamorphosis
- Suck blood of people and other vertebrates
- Pests: body lice, head lice, public lice.

Figure: *Order Mallophaga, a chewing louse*

- No wings
- Chewing mouthparts
- Broad head
- Feed on skin surfaces of birds and mammals
- Occasional human infestations.

Figure: *Order Orthoptera, a cockroach*

- Two pairs of wings or wingless
- Chewing mouthparts
- Simple metamorphosis
- Grasshoppers and crickets
- Pests: cockroaches.

Figure: *Order Hemiptera, a true bug*

- Two pairs of wings or wingless
- Top pair of wings partly leathery, partly transparent
- Piercing-sucking mouthparts
- Simple metamorphosis
- Pests: kissing bugs, bed bugs.

Figure: *Order Lepidoptera, a butterfly*

- Two pair of wings, usually scaled
- Sucking mouthparts
- Complete metamorphosis
- Damage many crops by feeding of caterpillars
- Rare instances of public health importance.

Figure: *Order Celeoptera, a beetle*

- Two pair of wings
- Outer pair hardened, inner pair soft
- Chewing mouthparts
- Complete metamorphosis
- Pests: blister beetles.

Figure: *Order Diptera, a house fly*

- One pair of wings
- Piercing-sucking or sponging mouthparts
- Complete metamorphosis
- Mosquitoes, flies, midges, and gnats are pests
- More species of public health importance than any other insect group.

Figure: *Order Hymenoptera, a wasp*

- Two pairs of wings
- Most adults with narrow waists
- Complete metamorphosis
- Ants are common pests
- Bees and wasps sting people and other animals.

Other Arthropods

Figure: *Subclass Acari, a tick*

- No distinct head
- Leathery or soft body
- No wings
- Parasitic on vertebrates
- Second only to mosquitoes in importance as vectors.

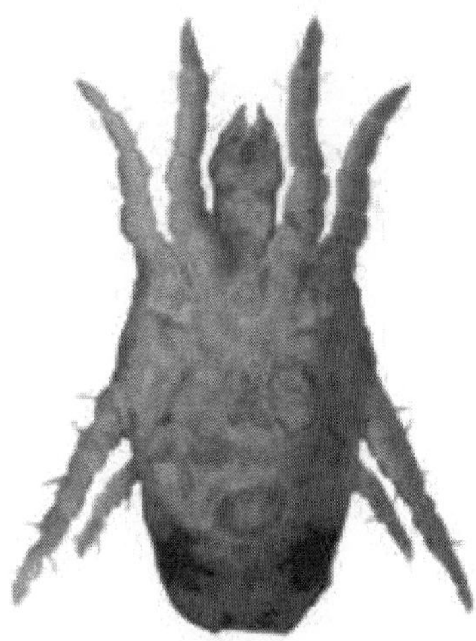

Figure: *Sublass Acari, a mite*

- Very small bodied
- Sucking mouthparts
- Parasites of plants and animals
- May transmit plant and animal pathogens.

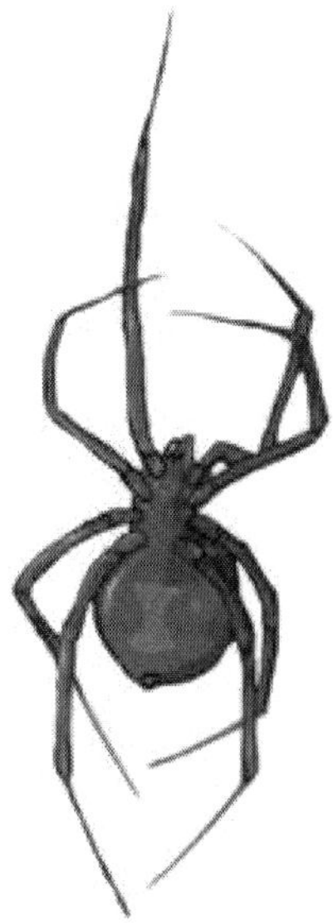

Figure: *Subclass Araneae, a spider*

- Fang-like mouthparts
- Most beneficial, a few poisonous
- Sucking mouth parts
- Parasitic on people and many other animals

Figure: *Subclass Scorpiones, a scorpion*

- Have stinging tail
- Some tropical species very dangerous
- Do not transmit pathogens
- Mainly pests in dry desert areas

Vertebrates

Vertebrates are animals with backbones. They include fish, frogs, toads, snakes, birds and mammals, including humans. Vertebrate pests are animals whose presence or activities are associated with human diseases or otherwise interfere with human interests and well-being. Introduced animals that alter the ecological balance in an area harming indigenous species may also be considered pest species. What may be a pest animal in some situations may be a highly desirable animal in others. For example, the beaver is considered a valuable fur-bearing animal; but because it can damage trees and create ponds that may serve as mosquito habitat it can also be a pest.

Common Vertebrates Pests

Fish

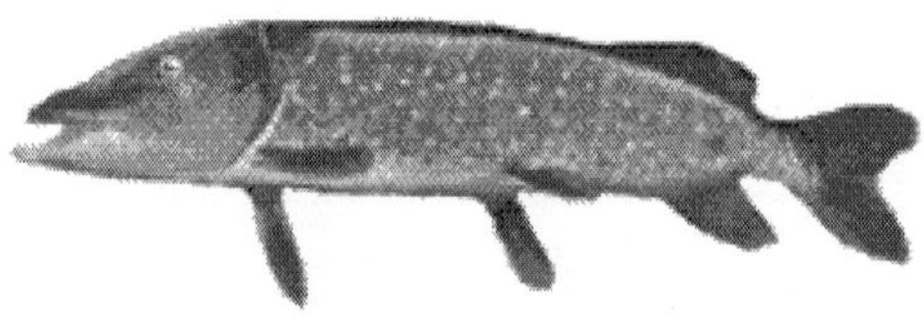

Figure: *A pike*

In California most fish pests are introduced species which compete with more desirable food and game fish. There are several invasive, non-native fish species in California that are considered pest species. The northern pike is such as species, and hundreds of millions of dollars are spent in California on efforts to remove it from lakes and reservoirs. Other fish species are undesirable because they serve as intermediate hosts for parasites of people and other vertebrates. In most areas of the state, mosquitofish are a valuable and important biological tool for vector control agencies, however there are locations where mosquito fish should not be used and may be considered pest. It is essential to know and follow state and federal law when introducing any non-native species.

Amphibians and Reptiles

Occasionally, toads or frogs may become nuisances around homes or parks. A few reptiles, in certain situations, might be considered pests. Most snakes are not poisonous, and they are generally beneficial. However, rattlesnakes are common in some areas of California, and when people come in contact with them there is a risk of receiving a poisonous bite. There are 6 species of rattlesnakes in California. The western diamond back rattlesnake is the most dangerous, and is responsible for most human deaths from snake bite in North America.

Even so, fatal snake bites are rare, with only 10-15 occurring each year in all of North America. This is far fewer than deaths caused by stings of bees and wasps.

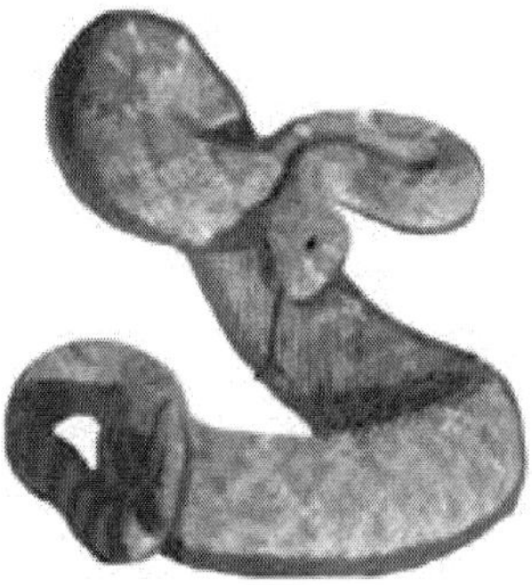

Figure: *A rattlesnake*

Birds

Many species of birds occur in California, many of which are beneficial and protected by law. However, birds can be serious pests and many species are involved in movement and amplification of human disease organisms. Most mosquito-borne viral diseases are principally diseases of birds, with humans being only secondarily infected.

Figure: *A crow*

Some species of birds can become pests in urban areas when they roost on buildings, leaving accumulations of droppings and otherwise causing damage. There also may be a risk of disease to humans who come in contact with droppings. In rural areas, birds may damage crops either by feeding directly on plant parts such as fruit, or by disrupting fields while feeding on seeds.

Control of bird pests requires specialized knowledge and techniques, and must be done in accordance with state and federal laws. Some vector control technicians are certified for vertebrate vector control and may work directly with birds. Most vector control technicians are not adequately versed in birds or the state and local laws to work directly with them. Anyone planning to undertake a project directed at birds should first consult his or her County Agricultural Commissioner and the California Department of Fish and Game.

Mammals

All mammals have warm blood, hair, and feed their young milk. Some mammal pests can represent serious potential risks to the health of humans and domestic animals. By biting, carnivorous mammals such as dogs, skunks, or coyotes may be vectors to humans of the virus causing rabies. Mammals may also serve as the primary hosts of several diseases that are transmitted (i.e., vectored) by arthropods.

Plague (transmitted by fleas) and tularemia (deer flies and ticks) are examples of such diseases, although humans can become infected directly by the bacteria causing the diseases in other ways as well. Hantavirus infection in people occurs by yet another route, where individuals typically become infected with the virus through contact with urine and feces of rodents.

Mammals can also be pests by killing livestock or damaging crops, lawns, stored goods, or structures.

Figure: *A coyote*

Weeds

A weed is any plant growing where it is not wanted. Because of this, weeds cannot be placed in any meaningful taxonomic classification. However, there are many non-native species that are legally considered weeds, despite their desirable qualities in flower gardens. The butterfly bush and Scotch Broom are prized by many gardeners in California because of its vigour and beautiful floral display, but others discourage the planting of this foreign species because of its tendency to grow out of control and crowd out native species.

Weeds can be damaging because they may:

- Compete with planted crops for water and nutrients, resulting in reduced crop yields, and impaired crop quality.
- Increase agricultural production costs.
- Interfere with proper operation of drainage and irrigation structures.
- Compete with indigenous plant species resulting in ecological damage and a corresponding loss of biological diversity.
- Degrade wildlife forage or hiding cover, or create habitat for species not previously in an area.
- Promote production of arthropod pests, especially mosquitoes, by providing protection from predacious fish and insects.
- Affect public health by producing allergens or causing contact dermatitis.

Common forms of Weeds

While there are about 500 plant species in California that generally are considered weeds, Most of these can be placed into one of two groups:

1. Narrow-leaved plants, including grasses, sedges, rushes, cattails, and a few others. The narrow-leaved weeds generally have parallel veins in the leaves.
2. Broad-leaved plants such as willow, oak, mustard, dock, pigweed, purslane, and bindweed. These weeds may be recognized by the presence of a net-like pattern of veins in the leaves. This is a useful classification for weed control purposes because these broad categories reflect not only a taxonomic separation (monocots and dicots) but are often important for the selection of an appropriate herbicide or other control strategy.

Weeds may be terrestrial or aquatic. Many aquatic weeds are important in mosquito abatement because they contribute to mosquito problems.

Life Cycles of Weeds

Plants also can be classified according to their life cycles as annuals, biennials, or perennials. Grouping weeds in this way can be helpful in designing weed control management plans. Annuals mature in one season and are propagated by seeds. Summer annuals grow from seeds that sprout in the spring, mature, and reproduce before dying in the winter. Among the common summer annuals are barnyard grass, puncture vine, Russian thistle, and pigweed. Winter annuals germinate in the fall or winter and grow until spring when they flower, produce seed, and die. Among the common winter annuals are chick-weed, mustard, wild oats, and annual bluegrass.

Biennials require two years to develop and complete their life cycles. These plants grow vegetatively (without flowering) through the first year and flower and produce seed during the second year. The plant then dies and will not re-appear at the site unless the seeds germinate and new plants grow. Common biennials are purple starthistle, musk thistle, mullein, and bristly ox tongue. Perennials are plants that live three years or longer.

These plants flower and set seed without dying. Perennials live indefinitely. They may die back to the ground in the winter, but re-grow in the spring. Common examples of perennials are Johnsongrass, field bindweed, dandelion, plantain, willow, and poison oak. Perennials are found throughout the Plant Kingdom, and they may take many forms. Some are herbaceous soft bodied plants, others are shrubs, and some are trees. Many perennials can store food underground in rhizomes, bulbs, or tubers, which makes these plants difficult to kill by mechanical or chemical means.

Weeds and Vector Control

Weed control in connection with vector control programs is limited to control of plants which may play a direct role in production of vector species and plants that prevent access to areas of vector production. Examples include overgrowth of cattails, tules, and bull rush. These plants can create areas of stagnant water and thick vegetation thus creating prime mosquito breeding habitat and making vector control difficult.

Pesticides as Hazardous Waste

Hazardous wastes are wastes with properties that make them dangerous or potentially harmful to human health or to the environment. Some pesticides are considered by EPA to fit this definition, and are thus subject to laws and regulations regarding their proper handling, transportation, and disposal.

Classification of Hazardous Wastes

A number of different federal and state agencies have enforcement responsibilities in regard to various aspects of hazardous waste management. The US Department of Transportation enforces regulations for movement of hazardous wastes. In 1976 the US Congress passed the Resource Conservation and Recovery Act (RCRA), which directed that EPA develop and implement a program to protect human health and the environment from improper hazardous waste management practices.

EPA has done this. In California, the Department of Toxic Substances Control (DTSC) is responsible for hazardous waste information and enforcement.

Listed Waste

EPA publishes lists of specific wastes that they consider hazardous. The lists are:

- F List (non-specific source wastes). This list consists of materials from common manufacturing and industrial processes such as cleaning solvents.
- K List (source-specific wastes). These are certain wastes from specific industries such as petroleum refining and pesticide manufacturing.
- P List and U-List (discarded commercial chemical products) include specific chemical products in an unused form. Some pesticides and pharmaceuticals become hazardous waste when discarded.
- M List (discarded mercury products; California). Included in this list are items such as fluorescent lamps and mercury switches.

Characteristic Waste

In addition to the EPA list, there are four criteria that are considered by them for chemical substances that may not be on one of their lists:

- Ignitability. These are substances than can create fires under certain conditions because they are spontaneously combustible, or have flash points less than 140*f*F. Waste oils and used solvents may meet this criterion.
- Corrosivity. Corrosive wastes are acids or bases (less than 2 or more than 12.5, respectively, that are capable of corroding metal tanks, drums, and barrels.
- Reactivity. Reactive wastes are substances that are unstable under normal conditions. They can cause explosions, toxic fumes, gases, or vapours when heated, compressed, or mixed with water.

- Toxicity. Toxic wastes are harmful or fatal when ingested or absorbed (e.g., those containing mercury or lead). There are 8 characteristics of toxicity defined in California, and if a waste meets even one of the 8, it is considered a toxic hazardous waste.

EPA has extensive information on hazardous substances on their website, listed above. Pesticide applicators should note that certain pesticides would fall under the category of hazardous waste based on several of the criteria listed here. Here is a general rule concerning what is and what is not a hazardous pesticide waste, aside from the criteria already covered: if the material in question contains any concentration of pesticide that can no longer be legally used, it must be considered a hazardous waste. If the material in question is considered no longer desirable as a pesticide, such as rinse material from containers and spray equipment, left-over spray solutions, excess pesticides, and cancelled/suspended pesticides, it must be disposed of as hazardous waste. If you are uncertain as to whether or not some quantity of a pesticide substance qualifies as a hazardous waste, you may:

1. Call the California DTSC at 916-327-4499
2. Send an email to the California DTSC: wasteclass@dtsc.ca.gov
3. Examine Material Safety Data Sheets for statements concerning it status as a hazardous waste
4. Read the pesticide label for information on its status
5. Write or call the formulator or distributor for information origin of hazardous wastes in vector control operations.

The most obvious origin of hazardous wastes related to vector control operations is pesticide applications. Some of the sources include empty pesticide containers containing hazardous residues, rinse water used to clean spray equipment and vehicles, pesticides remaining in sprayer tanks after an application, and unused stocks of outdated and possibly suspended pesticides. Another source of hazardous wastes is the repair and maintenance facility maintained by most mosquito and vector abatement agencies. In California, used motor oil is considered a hazardous waste. Solvents and other fluids used in connection with vehicle maintenance and repair may also fit the classification of hazardous.

One of the keys to management of hazardous waste is to examine the activities of an agency to determine which of them may be generating waste, and how much is generated on average per month.

Proper Management of Hazardous Wastes

Most mosquito and vector control agencies will generate waste that will qualify as hazardous. The US EPA has established categories for agencies that generate hazardous waste based on the amount of hazardous waste generated each month. These same categories have been adopted by the California DTSC.

It is important for agencies engaged in vector control operations to know the amount of waste they generate, because the rules for disposal of hazardous waste vary according to the category. This primarily applies to how long hazardous waste can be stored before it must be disposed of as there is an increased risk potential associated with storage of large quantities of these waste.

Table: *EPA Hazard Generator Categories*

Hazardous Waste Generated per Month	Category Name and Maximum Allowed Accumulation time
Less than 220 lbs (about half a drum)	CESQG (Conditionally exempt from requirement)
220 lbs –2,200 lbs (up to 5 drums)	SQG (Small Quantity Generator) 180 days (270 days if waste to be transported 200 miles or more
2,200 lbs or more (over 5 drums)	LQG (Large Quantity Generator) 90 days

Containers Used to Collect Hazardous Wastes

Hazardous materials can be placed in their original containers temporarily, but the most common containers used to store hazardous wastes prior to their disposal are 55-gallon steel or plastic drums and inner liners from these drums.

If you have dedicated containers for pesticide hazardous waste, they must contain the following label:

Hazardous Waste

Federal Law Prohibits Improper Disposal

If found, please contact the nearest police, public safety authority, or the US EPA. Other information that should be placed on waste containers include the date waste was first put in the container, the federal waste code numbers, and the type of waste contained in the container. Even if the waste is not classified as hazardous waste, it is a good practice to label the container with the waste material it contains.

Empty Containers

Regulations that address the management of empty pesticide containers are very complex. The DTSC has prepared a fact sheet "Managing Empty Containers" that contains an extensive question and answer section. This fact sheet is available on their website. California regulations recognize two categories of containers that are exempt from management rules for hazardous waste:

(1) containers that once contained waste that does not qualify as hazardous, and

(2) containers that meet the definition of "empty" and have been subjected to approved management practices.

Containers that are exempt can be disposed of at any appropriate solid waste facility.

From the standpoint of management of hazardous wastes, "empty" containers are defined as containers that once held hazardous materials or hazardous wastes and have been emptied by the generator (in most cases the vector control agency that used the pesticide) as much as is reasonably possible. In practice, this means inverting the container until no more liquid or solid material pours out.

However, because of the concern that containers that fit the definition of "empty" may still contain some residual hazardous materials that could cause significant harm, they are still not considered non-hazardous unless approved management practices are followed. Containers that held liquid pesticides classified as acute or extremely hazardous waste must be triple-rinsed with a solvent capable of removing the material before they can be considered ready for disposal. It would be unusual for most vector control agencies to use pesticides fitting the latter category.

Containers that held solid pesticides, or pesticides that have become viscous and pour slowly, or have dried out or otherwise become caked, must be cleaned using various methods, such as scraping, in addition to rinsing to remove all pesticide residues. Management practices required are flexible, depending upon the way empty containers are handled by a particular agency. Some may return empty containers to the original distributor; others may ship containers to a company that reconditions them. Empty containers of 5 gallons or less can be disposed of at a solid waste facility.

Large Containers

Some mosquito and vector control agencies use special underground tanks to collect hazardous wastes such as rinse water. Underground storage tanks are not considered containers, and their management is covered under other regulations. Bulk containers of 110 gallons or more also are controlled by a different set of regulations.

Storage of Containers Full or Partially Full of Pesticides

Containers of pesticides over time can deteriorate to the point where they can rupture, leak, rust, and otherwise lose their original integrity. Especially if pesticides are stored over long periods of time (years) they should be periodically inspected for signs of deterioration, and if leaking containers are found, their contents should be transferred to a new container.

Containers should be kept closed at all times. If containers are used to contain waste, self-closing funnels should be used to add waste. Waste pesticides should never be allowed to evaporate.

Pesticides should always be compatible with the material the container is made of. If the waste is corrosive, high density plastic containers should be used. Wastes that are reactive (e.g., acids and bases) should never be kept in the same container.

Containers for waste pesticides should be stored indoors. Adequate aisle space and ventilation should be provided. This makes periodic inspections for leaks and container deterioration easier. Ignitable and reactive pesticide wastes should be stored at least 50 feet from property boundaries. Pesticide wastes, depending upon the amount of waste generated per month, have legal time limits they can be stored before disposal.

Required Notifications of Hazardous Wastes

Agencies that generate pesticide hazardous waste should designate an emergency coordinator. This person must know what to do in case of a poisoning, fire, spill, or other emergency, and should be available 24 hours a day, 7 days a week.

The following information should be clearly posted in a prominent place, such as near a telephone in the main district office, or near a telephone in the proximity of the storage site for pesticide hazardous wastes:

- Local fire department number
- Emergency Coordinatorís name and telephone number
- Locations of fire alarms and fire extinguishers
- Location of pesticide spill control materials.

Pesticides not Classified as Hazardous Wastes

- May be disposed of as regular solid waste or trash.
- Are regulated by California law
- Must be disposed of according to instructions contained on the product label
- Must be disposed of in a careful manner.

Proper Transportation of Hazardous Wastes

In California the transportation of hazardous wastes are regulated by the California Department of Toxic Substances Control (DTSC), and no hazardous wastes may be transported without first getting a permit from DTSC. Because of the complex and expensive requirements for permission to transport hazardous wastes, most mosquito and vector control districts contract with a licensed operator to remove, transport, and dispose of hazardous wastes.

Those who believe they would like to transport and dispose of hazardous wastes in-house should refer to "Hazardous Waste Transporter Requirements", August 2007, available on the DTSC website. After reading the daunting requirements, few pesticide technicians will want to pursue this route. Pesticide technicians should know that any person who transports hazardous wastes must hold a valid registration issued by DTSC. Further, it is unlawful for any person to transfer custody of toxic waste to a transporter who does hold a valid registration. If in doubt, there is a list of registered hazardous waste transporters on the DTSC website.

Proper Disposal of Hazardous wastes

This section should be entitled "How to avoid having to dispose of hazardous wastes". As mentioned in the previous section, the actual disposal of hazardous wastes is in the most part handled by commercial firms that specialize in this area. Recall also that agencies that generate less than a threshold amount of hazardous waste in a month are exempt from many of the management requirements.

Here are some practices used by many California mosquito and vector control agencies to minimize the amount of hazardous waste that requires special handling and management:

- Keep the use of pesticides that qualify as hazardous waste to a minimum.
- Avoid the use of pesticides that are on the Acutely Toxic Hazardous Waste list of EPA. Nearly all of these products are no longer available for use in California anyway.

- As much as possible, completely empty sprayer tanks of pesticide during applications. Avoid coming back from a job with partially-filled tanks.
- Do not collect rinse water from pesticide equipment cleaning in tanks for long-term storage, but use the rinse water for diluting water-soluble pesticides in spray tanks. This can only be done if the pesticide sprayed in the equipment that was cleaned is compatible with the pesticide to be diluted with the rinse water.
- Take advantage of recycling programs for empty pesticide containers offered by many pesticide distributors.

Chapter 7

Toxicity of Pesticides

Introduction

This chapter describes how pesticides work on living things, so you can learn to reduce or eliminate exposure. Pesticides are designed to control pests, but they can also be toxic (poisonous) to desirable plants and animals, including humans. Some pesticides are so highly toxic that very small quantities can kill a person; while exposure to a sufficient amount of almost any pesticide can make a person ill.

Since even fairly "safe" pesticides can irritate the skin, eyes, nose, or mouth, it is a good idea to understand how pesticides can be toxic so you can follow practices designed to reduce or eliminate your exposure and the exposure of others to them.

How Pesticides Enter the Body

Before a pesticide can harm you it must be taken into the body. Pesticides can enter the body orally (through the mouth and digestive system); dermally (through the skin) or by inhalation (through the nose and respiratory system).

Oral Exposure

Oral exposure may occur because of an accident, but is more likely to occur as the result of carelessness, such as blowing out a plugged nozzle with your mouth, smoking or eating without washing your hands after using a pesticide, or eating fruit that has been recently sprayed with a pesticide.

The seriousness of the exposure depends upon the oral toxicity of the material and the amount swallowed.

Dermal Exposure

Dermal (skin) exposure accounts for about 90% of the exposure pesticide users receive from nonfumigant pesticides. It may occur any time a pesticide is mixed, applied, or handled, and it often goes undetected. Dry materials—dusts, wettable powders, and granules, as well as liquid pesticides—can be absorbed through the skin. The seriousness of dermal exposure depends upon:

- The dermal toxicity of the pesticide;
- Rate of absorption through the skin;
- The size of the skin area contaminated;
- The length of time the material is in contact with the skin; and
- The amount of pesticide on the skin.

Rates of absorption through the skin are different for different parts of the body. Using absorption through the forearm as the standard, absorption is over 11 times faster in the lower groin area than on the forearm. (Absorption through the skin in the scrotal area is rapid enough to approximate the effect of injecting the pesticide directly into the bloodstream.)

Absorption continues to take place on all of the affected skin area as long as the pesticide is in contact with the skin. The seriousness of the exposure is increased if the contaminated area is large or if the material remains on the skin for a period of time.

Table : *Parathion absorption rates through the skin on various bodily regions.*

Body region	*Percent relative absorption*
Forearm	8.6
Palm of hands	11.8
Ball of foot	13.5
Abdomen	18.4
Scalp	32.1
Forehead	36.3
Ear canal	46.5
Genitalia	100

Inhalation Exposure

Inhalation exposure results from breathing pesticide vapours, dust, or spray particles. Like oral and dermal exposure, inhalation exposure is more serious with some pesticides than with others, particularly fumigant pesticides. Inhalation exposure can occur from the applicator's smoking; breathing smoke from burning containers; breathing fumes from pesticides while applying them without protective equipment; and inhaling fumes while mixing and pouring pesticides.

Toxicity

Toxicity refers to the ability of a poison to produce adverse effects. These adverse effects may range from slight symptoms such as headaches to severe symptoms like coma, convulsions, or death. Poisons work by altering normal body functions, consequently toxicity can occur in as many ways as there are body functions. Most toxic effects are reversible and do not cause permanent damage if prompt medical treatment is sought. Some poisons, however, cause irreversible (permanent) damage. All new pesticides are tested to establish the type of toxicity and the dose necessary to produce a measurable toxic reaction. In order to compare the results of toxicity tests done in different labs, there are strict testing procedures. Toxicity testing is extensive (involving many phases) and therefore, expensive. Humans, obviously, cannot be used as test animals so toxicity testing is done with animals. Since different species of animals respond differently to chemicals, a new chemical is generally tested in mice, rats, rabbits, and dogs. The results of these toxicity tests are used to predict the safety of the new chemical to humans.

Toxicity tests are based on two premises. The first premise is that information about toxicity in animals can be used to predict toxicity in humans. Years of experience have shown that toxicity data obtained from a number of animal species can be useful in predicting human toxicity, while data obtained from a single species may be inaccurate. The second premise is that by exposing animals to large doses of a chemical for short periods of time, we can predict human toxicity from to exposure to small doses for long periods of time. Toxicity is usually divided into two types, acute or chronic, based on the number of exposures to a poison and the time it takes for toxic symptoms to develop. Acute toxicity is due to short-term exposure and happens within a relatively short period of time, whereas chronic exposure is due to repeated or long-term exposure and happens over a longer period..

Acute Toxicity

The acute toxicity of a chemical refers to its ability to do systemic damage as a result of a one-time exposure to relatively large amounts of the chemical. A pesticide with a high acute toxicity may be deadly if even a very small amount is absorbed. The signal words on the label are based on the acute toxicity of the pesticide. Acute toxicity may be measured as acute oral (through the mouth), acute dermal (through the skin) and acute inhalation (through the lungs or respiratory system).

Table: *Types of Toxicity*

Type	Number of Exposures	Time for symptoms to develop
Acute	usually 1	immediate (minutes to hours)
Chronic	more than a few	one week to years

Table: *Acute Toxicity Measures and Warnings*

		Categories of Acute Toxicity			
LD_{50}	LD_{50}	LC_{50}			
Categories	Signal Word	Oral mg/kg	Dermal mg/kg	Inhale mg/l	Oral Lethal Dose[1]
I Highly Toxic	DANGER, POISON (skull & crossbones)	0 to 50	0 to 200	0 to 2,000	a few drops to a teaspoonful
II Moderately Toxic	WARNING	50 to 500	200 to 2,000	2,000 to 20,000	over a teaspoonful to one ounce
III Slightly Toxic	CAUTION	500 to 5,000	2,000 to 20,000	n/a	over one ounce to one pint
IV Relatively Non-toxic	CAUTION	5,000+	20,000 +	n/a	over one pint to one pound

[1] Probable for a 150 lb.-person.

Acute Toxicity Measure

The commonly used term to describe acute toxicity is $LD_{50.}$ LD means Lethal Dose (deadly amount) and the subscript 50 means that the dose was acutely lethal to 50 percent of the animals to whom the chemical was administered under controlled laboratory conditions. The test animals are given specific amounts of the chemical in either one oral dose or by a single injection, and are then observed for 14 days.

Since LD_{50} values are measured from zero up, the lower the LD_{50} the more acutely toxic the pesticide. Therefore, a pesticide with an oral LD_{50} of 500 would be much less toxic than a pesticide with an LD_{50} of 5. LD_{50} values are expressed as milligrams per kilogram (mg/kg) which means milligrams of chemical per kilogram of body weight of the animal. Milligram (mg) and kilogram (kg) are metric units of weight similar to ounce and ton. Milligrams per kilogram is the same as parts per million.

For example, if the oral LD_{50} of the insecticide parathion is 4, a dose of 4 parts of parathion for every million parts of body weight would be lethal to at least half of the test animals.

LD_{50} values are generally expressed on the basis of active ingredient. If a commercial product is formulated to contain 50 percent active ingredient, it would take two parts of the material to make one part of the active ingredient.

In some cases, other chemicals mixed with the active ingredient for formulating the pesticide product may cause the toxicity to differ from that of the active ingredient alone. Acute inhalation toxicity is measured by $LC_{50.}$ LC means Lethal Concentration. Concentration is used instead of dose because the amount of pesticide inhaled from the air is being measured. LC_{50} values are measured in milligrams per litter. Liters are metric units of volume similar to a quart. The lower the LC_{50} value, the more poisonous the pesticide.

Chronic Toxicity

Chronic toxicity refers to harmful effects produced by long-term, low-level exposure to pesticides. Less is known about the chronic toxicity of pesticides than is known about their acute toxicity, not because it is of less importance, but because chronic toxicity is much more complex and subtle in how it presents itself.

While situations resulting in acute exposure (a single large exposure) do occur, they are nearly always the result of an accident or careless handling. On the other hand, persons may be routinely exposed to small amounts of pesticides while mixing, loading, and applying pesticides or by working in fields after pesticides have been applied.

Chronic Toxicity Measures

There is no standard measure like the LD_{50} for chronic toxicity. How chronic toxicity of chemicals is studied depends upon the adverse effect being studied. Chronic adverse affects may include carcinogenesis, teratogenesis, mutagenesis and reproductive toxicity.

Carcinogenesis (Oncogenesis)

Carcinogenesis means the production of malignant tumours. Oncogenesis is a generic term meaning the production of tumours which may or may not be carcinogenic. The terms tumour, cancer, or neoplasm are all used to mean an uncontrolled progressive growth of cells. In medical terminology, a cancer is considered a malignant (potentially lethal) neoplasm.

Carcinogenic or oncogenic substances are substances which can cause the production of tumours. Examples are asbestos and cigarette smoke.

Teratogenesis

Teratogenesis is the production of birth defects. A teratogen is anything that is capable of producing changes in the structure or function of the offspring when the embryo or fetus is exposed before birth. An example of a chemical teratogen is the drug thalidomide which caused birth defects in children when their mothers used it during their pregnancy. Measles virus infection during pregnancy has teratogenic effects.

Mutagenesis

Mutagenesis is the production of changes in genetic structure. A mutagen is a substance which causes a genetic change. Many mutagenic substances are oncogenic, meaning they also produce tumours. Many oncogenic substances are also mutagens.

Reproductive Toxicity

Some chemicals have effects on the fertility or reproductive rates of animals. Males or females can be affected.

Chronic Toxicity Testing

Chronic toxicity testing is both lengthy and expensive. EPA and regulatory agencies in other countries require an extensive battery of tests to identify and evaluate the chronic effects of pesticides. These studies, which may last up to 2 years, utilize several species of animals to evaluate toxicity from multiple exposures or continuous long-term exposure.

Label Identification of Acute and Chronic Toxicity

To alert pesticide users to the acute toxicity of a pesticide, a signal word must appear on the label. Four different categories are used. Signal words are used to tell the user whether the chemical is highly toxic, moderately toxic, slightly toxic, or relatively non-toxic. These label warnings are based, for the most part, on the chemical's acute toxicity. For example, the acute oral and acute dermal toxicity of a pesticide may be in the slightly toxic category. But if the acute inhalation toxicity is in the highly toxic category, the pesticide label will have the signal words for a highly toxic pesticide. The degree of eye or skin irritation caused by the pesticide also influences the signal word.

For chronic toxicity there is no comparable set of signal words like those used for acute toxicity. Instead, a statement identifying the specific chronic toxicity problem is sometimes used on the label. Such a statement might read "This product contains (name of chemical) which has been determined to cause tumours or birth defects in laboratory animals." Chronic toxicity warning statements may be accompanied by label directions to wear certain kinds of protective clothing when handling or working with the pesticide to minimize or eliminate exposure to the pesticide.

It is important to read the label to look for signal words identifying the product's acute toxicity and for statements identifying any chronic toxicity problem. A pesticide may be low in acute toxicity (signal word caution), but it may have a label statement identifying potential chronic toxicity.

Safety Factors

In animal feeding studies, the pesticide under investigation is incorporated into the daily diet and fed to animals from a very young to a very old age. These, as well as the reproductive effects studies, are designed to arrive at a No-Observable-Effect-Level (NOEL); that is, a level in the total diet that causes no effect in treated animals when compared to untreated animals maintained under identical conditions. This NOEL is expressed on a mg/kg of body weight/day basis. A Reference Dose (RfD), also known as Acceptable Daily Intake (ADD), is usually established at 1/100 of the NOEL, in order to add an additional margin of safety. The RfD (ADI) is the amount of chemical that can be consumed daily for a lifetime without ill effects.

Extensive residue trials are conducted to determine levels of the pesticide that remain in or on growing crops after treatment with the pesticide. These trials lead to the establishment of a tolerance for residues of the chemical on food commodities. A tolerance is the maximum allowable amount of the pesticide permitted in or on a specific food commodity at harvest. The directions for use found on the product label are written to assure that residues in food commodities are below the tolerance. The tolerance is set low enough to assure that even if someone ate only food items with residues of a given pesticide at the tolerance limit, there would still be a safety factor of at least 100 when compared to a level causing no observable effects in laboratory animals. This is, of course, a worst case situation, since all crops on which the pesticide is registered for use would not be treated with the chemical, and in most cases residue levels would be well below the tolerance due to pre-harvest intervals being longer than the minimum period stated on the label. Further reduction of residues may occur in storage or from washing, trimming, and processing.

Dose-response

Ironically, the extensive amount of data developed about a pesticide is often used against it by ignoring the dose-response. For example, some acute toxicity studies, which are designed to include dosage levels high enough to produce deaths, are cited as proof of the chemical's dangers. Chronic effects seen at very high doses in lifetime feeding studies are misinterpreted and considered as proof that no exposure to the chemical should be allowed. Major improvements in analytical chemistry permit detection of the presence of chemicals at levels of parts per billion (ppb) or even parts per trillion (ppt).

We may hear that a certain chemical has been found in a food or beverage and the amount found is expressed in parts per million or parts per billion. Often, no information is provided to assist us in comprehending the meaning of these numbers. Frequently, this information neglects the issue of dose-response, the key principle of toxicology which simply stated is "the dose makes the poison.". The concentration of a chemical in any substance is meaningless unless it is related to the toxicity of the chemical and the potential for exposure and absorption. Chemicals of low toxicity such as table salt or ethyl alcohol can be fatal if consumed in large amounts. Conversely a highly toxic material may pose no hazard when exposure is minimal.

Monitoring for Residues

Monitoring foods for pesticide residues is carried out by the Food and Drug Administration (FDA) and the U.S. Department of Agriculture (USDA). Crops containing residues over the official tolerance (maximum legal level) established by the EPA must be destroyed. The threat of crop destruction with resultant financial loss is a strong incentive for farmers to observe use instructions on pesticide labels and thus assures that residues will be below established tolerances. Crops grown for export are often checked for residues by foreign residue laboratories to assure that local tolerance limits are not exceeded. Lastly, market-basket surveys (analyses of food items from grocery stores) have confirmed the low exposure of the general public to pesticides in foods.

Hazard

Hazard is a function of the toxicity of a pesticide and the potential for exposure to it. We do not have control of the toxicity of a pesticide since toxicity is a given characteristic of a particular pesticide; however, we can have control over our exposure to pesticides. This is done by following several safety practices including the use of protective clothing and equipment (PPE). All pesticides are hazardous if misused, no matter what their toxicity. All pesticides can be handled safely by using safety practices that minimize or eliminate exposure to them.

Federal laws regulating pesticides have placed the burden of proving safety of pesticide usage on the manufacturer. Hazard evaluation studies are generally done by scientific laboratories maintained by the manufacturer or through outside contract laboratories. Few products are subjected to such extensive and vigorous testing as pesticides are before they are marketed. Many promising pesticide products are not marketed because they do not pass the extensive toxicology testing. Older pesticide products that were registered before the current toxicology testing standards were established are being re-evaluated under current standards. Precautions and other safety information found on the product's label are based on information from these tests. By reading and following the directions on the label, users can minimize or eliminate hazards due to use of the pesticide to themselves and others.

Common Pesticide Poisonings

The pesticides most often implicated in poisonings, injuries, and illnesses, according to 1996 data from the American Association of Poison Control Centre's Toxic Exposure Surveillance System, are listed in Table 4. The list is based on symptomatic cases classified as minor, moderate, major, or fatal outcome for unintentional cases involving a single product. Numbers of cases are reported for both children under six years of age and for adults and older children. Suicide/homicide (intentional) cases have been excluded. Cases listed as organophosphates (and the other categories as well) may also include other insecticides such as carbamates and organochlorines in a single product.

Approximately 90% of symptomatic cases involve only minor symptoms of the type that could typically be treated at home with dilution or just observation. However, according to the American Association of Poison Control Centres, seven of the top ten categories listed in the table (organophosphates, pyrethrins/pyrethroids, hypochlorite disinfectants, carbamates, organochlorines, phenoxy herbicides, and anticoagulant rodenticides) are much more likely to require medical attention.

This list can't be considered representative of all symptomatic poisonings because it only shows cases reported to Poison Control Centres.

However, it does give a sense of the relative frequency and risk of poisoning from various agents or classes of agents. The relative frequency of cases generally reflects how widely a product is used in the environment.

For example, a number of disinfectants occur in the top ten partly because they are far more commonly found in the home and work environment than other pesticides. Denominator information on the population at risk (numbers exposed) would be needed to better understand the relative risk of different pesticides.

However, the main purpose is to give physicians a sense of what types of cases they are most likely to see in their practice.

Table: *Pesticides most often implicated in symptomatic illnesses, 1996.**

Rank	*Pesticide or Pesticide Class*	*Child <6 years*	*Adults and those 6 – 19 years*	*Total**
1	Organophosphates	700	3,274	4,002
2	Pyrethrins and pyrethroids**	1,100	2,850	3,950
3	Pine oil disinfectants	1,336	903	2,246
4	Hypochlorite disinfectants	808	1,291	2,109
5	Insect repellents	1,081	997	2,086
6	Phenol disinfectants	630	405	1,040
7	Carbamate insecticides	202	817	1,030
8	Organochlorine insecticides	229	454	685
9	Phenoxy herbicides	63	387	453
10	Anticoagulant rodenticides	176	33	209
	All other pesticides	954	3,604	4,623
Total all pesticides and disinfectants		7,279	15,015	22,433

*Totals include a small number of cases with unknown age.

**Rough estimate: includes some veterinary products not classified by chemical type.

Source: American Association of Poison Control Centres, Toxic Exposure Surveillance System, 1996 data.

General Classification of Pesticides: Fumigants

Fumigants are The Most Toxic of all pesticide chemicals used in agriculture. They are used to kill insects, nematodes, weed seeds and fungi in soil, silo-stored cereal grains, fruits and vegetables. Fumigants may be liquids (ethylene dibromide, dibromochloropropane, formaldehyde, etc) that readily vaporise at ambient temperature, solid that can release a toxic gas on reacting with water or acid, or gases (methyl bromide, hydrogen cyanide, ethylene oxide, etc). They are highly biologically reactive compounds.

Many are alkylating agents, mutagens, carcinogens, neurotoxins and hepatotoxins. These chemicals are gases penetrating readily into the lungs where they enter the blood and are rapidly distributed throughout the body. Full respiratory protection is mandatory when working with these pesticides. Death could occur rapidly if the fumes are inhaled, even for a short period of time. However, skin absorption can be significant hazard as well. Fumigants, as a class, have caused severe human illness and death.

Methyl Bromide

Methyl bromide is a widely used soil sterilant and post-harvest fumigant in agriculture against insects, termites, rodents, weeds, nematodes and soil-borne diseases. It is used to fumigate agricultural commodities, grain elevators, mills, ships, clothes, furniture and greenhouses. Non-pesticidal uses include de-greasing wool and extracting oils from nuts, seeds and flowers. Methyl bromide is also a methylating agent in the chemical industry. In the past, it was used as a refrigerant and as a fire-extinguishing agent in aircraft. The chemical name for this fumigant is bromomethane.

Methylbromide, labelled with the word "DANGER", is an extremely toxic vapour. It is a colourless fas with a burning taste and sweet chloroform-like odour at high concentrations. It is non-flammable in air but does burn in oxygen. It is stable and non-corrosive. In humans, methyl bromide is readily absorbed through the lungs. Most problems occur as a result of inhalation. Methyl bromide is a dangerous cumulative poison. First symptoms often are due to damage to the nervous system and may be delayed from 48 hours to as long as severak months after exposure. This delay, combined with methyl bromide's lack of odour, means that the victim may not realise that exposure is occurring until much time has passed. Symptoms of poisoning vary widely. Soon after inhalation of large doses, symptoms may include headache, dizziness, nausea, chest and abdominal pain and a dry throat. Three to 12 hours after vapour inhalation, symptoms include slurred speech, blurred vision, temporary blindness, mental confusion and sweating. More severe symptoms may include lung swelling, congestion, haemorrhaging of the brain, heart and spleen, severe kidney damage and numbness. Death may occur within one to 30 hours, usually from respiratory failure.

Although skin absorption is not an important route foe methyl bromide intocication, the skin is affected by contact with this chemical. methyl bromide can cause enormous blisters that are rarely deep enough to destroy the entire skin layer. Small amounts of skin or eye contact brings on shortness of breath and itching. If absorbed through skin, nausea and vomiting may result. Clothing that cannot "breathe" may delay the evaporation of the pesticide from the skin. Continued contact with skin can cause death. Ingestion of methyl bromide may cause hand tremors and convulsions. Chronic exposure to methyl bromide can cause dizziness, vision and hearing disturbances, depression, confusion, hallucinations, euphoria, personality changes and irritability. If exposure is severe enough, lung irritation, followed by lung swelling and bronchial pneumonia may occur.

Hydrogen Cyanide

Hydrogen cyanide is an insecticidal fumigant used mainly for stored products, especially grains and flour in mill, warehouses and ships. In the United States, it has been used extensively for combating scale insects on citrus trees while each tree is covered by a gas-proof tent. It is colourless gas or liquid with a characteristic almond-like odour. It burns with a blue flame. Hydrogen cyanide may be absorbed in toxic amounts through the unbroken skin (the amount is increased if the skin is moist), inhalation of gas and by ingestion. The cyanide ion has essentially the same toxicity, regardless of the route by which it is absorbed. However, accidental or intentional poisoning (excluding executions) is more likely to involve ingestion of a cyanide salt whereas occupational poisoning is more likely to involve inhalation of hydrogen cyanide, whether it is volatilised from liquid or is generated at the site of use by the actionof a mineral acid on potassium or sodium cyanate. A large doses of cyanide, whether oral or respiratory, is followed by almost instantaneous collapse and cessation of respiration. A dosage only a little above the threshold of danger causes only headaches, weakness, confusion and sometimes nausea and vomiting. respiration and heart rate are stimulated at first, but respiration later becomes slow and gasping.

The heart may continue for some time after respiration stops. While respiration is active, the venous blood remains oxygenated and the patient's colour florid. In fact, this condition may persist if death is sudden. If respiration is extremely poor or if it stops some moments before the heart, the blood will become dark and the patient cyanotic. The toxic action is reversible and a person who is completely unconscious from the effects of cyanide but whose heart is still beating, may still recover if suitable antidotes and remedial measures are applied on time.

Ethylene dibromide

Ethylene dibromide has been used as a fumigant for stored grain, fruits and vegetables for the treatment of soil against nematodes and for spot treatment in flour mills. The compound is sometimes mixed with carbon tetrachloride or dicholorpropene to increase the spectrum of activity. It is a heavy, colourless liquid with a chloroform-like odour. It is stable and non-flammable. Ethylene dibromide is a severe irritant to skin, eyes and the respiratory tract.

The liquid causes blistering and erosion of the skin and is corrosive to the eyes. Once absorbed, it may cause pulmonary oedema and central nervous system depression. Long-term exposure may have some damaging effect on testicular tissue. Persons poisoned by ingestion have suffered chemical gastroentetitis, liver necrosis and renal tubular damage. Death is usually due to respiratory or circulatory failure. A powerful disagreeable odour is advantageous in warning occupationally exposed workers of the presence of this gas.

Dibromochloropropane

Dibromochloropropane is used as a soil fumigant and nematocide. It forms a brown liquid with a pungent odour. Dibromodichloropropane is irritating to the skin, eyes and the respiratory tract. Eye damage has resulted from repeated exposure to the vapours.

When absorbed, it causes headache, nausea, vomiting, ataxia and slurred speech. Liver and kidney damage are prominent features of acure poisoning. Chronic exposure to relatively low concentrations has led to permanent sterility of workers in a mnaufacturing plant. Because it is much less odiferous then ethylene dibromide, exposure of workers to toxic concentrations of dibromochloropropane is more likely.

Ethylene Oxide and Propylene Oxide

Ethylene oxide and propylene oxide are irritants to all tissues they come in contact with. Aqueous solutions of ethylene oxide cause blistering and erosion of the affected skin. The area of the skin may thereafter be sensitised to the fumigant. Inhalation of high concentrations is likely to cause pulmonary oedema and cardiac arrhythmias. Headache, nausea, vomiting, weakness and a persistent cough are common early manifestations of acute poisoning. Coughing of bloody frothy sputum is characteristic of pulmonary oedema.

Phospine Gas

Phosphine is a fumigant for grain and other stored products. The fas is liberaed by the interaction of aluminium phosphide with water. Phosphine is a colourless gas with an odour of carbide or decaying fish. Phosphine in the form of zinc phosphide is also used in the control of rodents and moles. Phosphine gas is only slightly irritating to the respiratory tract but is at least as toxic systematically as hydrogen cyanide. It is slowly released into treated produce or storage spaces by hydrolysis of solid aluminium phosphide. Mechanisms of toxicity are not well understood. The principle manifestations of poisoning are fatigue, anusea, headache, dizziness, thirst, cough, shortness of breath, paraethesia and jaundice. Pulmonary oedema is a common cause of death.

What are Pesticides?

The *International code of conduct on the distribution and use of pesticides* defines pesticides as: "Any substance or mixture of substances intended for preventing, destroying or controlling any pest, including vectors of human and animal disease, unwanted species of plants or animals causing harm during, or otherwise interfering with, the production, processing, storage, transport, or marketing of food, agricultural commodities, wood and wood products or animal foodstuffs, or which may be administered to animals for the control of insects, arachnids or other pests in or on their bodies. The term includes substances intended for use as a plant growth regulator, defoliant, desiccant, or agent for thinning fruit or preventing the premature fall of fruit, and substances applied to crops either before or after harvest to protect the commodity from deterioration during storage and transport."

The Problem

All pesticides are toxic to some or all living organisms. They are designed to prevent, destroy or control specific plants or animals that threaten crops or other useful resources. However, if beneficial insects or crops are exposed to pesticides they too may be destroyed, and farm animals, wildlife or people may become ill or die after exposure to even very small quantities of pesticide. Over the past 40 years, the worldwide production and use of pesticides have increased.

In 1996, the global pesticide market was valued at US$30 560 million. Growth in pesticide sales has slowed in industrialized countries, but it continues to grow rapidly in developing countries and dependence on pesticides is also increasing in these regions. As reliance on pesticides increases, so do the problems of hazardous pesticide wastes. How can used containers, contaminated materials, deteriorated or unusable chemicals and excess pesticide supplies be disposed of safely?

The World Health Organization (WHO) estimates that, worldwide, exposure to pesticides causes an annual 20 000 deaths and at least 3 million cases of acute poisoning. Other estimates suggest that the annual figure for pesticide poisonings is as high as 25 million in developing countries alone. Many deaths and cases of poisoning are caused by mishandling of pesticide wastes and containers; the common practice of reusing pesticide containers to store food and water is an example of this. Pesticides that are carelessly disposed of can contaminate the air, water and land, and poison people, livestock, fish and wildlife.

Size of the Problem

Commercial synthetic chemical pesticides are traded in every country and reach even the remotest regions. Wherever pesticides are used, unusable or unwanted pesticides and empty pesticide containers have to be managed and disposed of safely. A wide variety of media (pictograms, posters, radio programmes and training of extension services and farmers) have been used to promote the safe storage, transportation and use of pesticides.

The success of such efforts varies, but their focus is on the safe use of pesticides, while relatively little has been done to promote the safe management of pesticide-related waste materials. Very few local or national authorities, pesticide suppliers or other organizations provide adequate support to pesticide end users. Advice is sometimes given on product labels or through regulations or guidelines, but this advice rarely reaches pesticide end users and, even when it does, it tends to be inappropriate and difficult to follow. People often reuse empty plastic or metal pesticide containers as storage for fuel or even food and water, even though it is usually impossible to remove all traces of chemicals from these containers.

Old products, unlabelled containers and leaking packages are often kept in misguided efforts to avoid waste or because people do not understand the hazards they pose. Other dangers arise when unwanted pesticides and containers are disposed of inappropriately. For example, many pesticide suppliers and national authorities recommend the burying or burning of waste pesticides and empty containers. However, buried chemical waste can contaminate soil and leach into surface or ground-water, while burning pesticides and containers release highly toxic fumes.

The Solutions

Pesticide regulators, advisers, distributors and end users urgently need clear advice on how to minimize the accumulation of pesticide-related waste and how best to deal with any waste that is generated. At present, guidance is often unclear, wrong or, in too many cases, completely lacking. Advice should focus, in particular, on ways in which pesticide end users can safely dispose of empty pesticide containers and unwanted or obsolete pesticides.

The main aim of these guidelines is to indicate best practices and responsibilities. The safe management and disposal of pesticide-related waste should be provided and coordinated by regulatory authorities, pesticide distributors and suppliers. Other organizations that support and advise pesticide users, such as extension and health promotion services, non-governmental organizations (NGOs), agricultural colleges and schools, also have important roles to play. Governments and their agencies, including ministries of agriculture, health, environment and education, are responsible for regulating the manufacture, import, distribution and use of pesticides. These responsibilities should be extended to include the management of pesticide-related waste products, including empty containers, which are often overlooked.

Pesticide suppliers should ensure that effective product stewardship systems are in place. Stewardship should cover all stages of pesticide production, distribution and use, including the management of waste. Local distributors of pesticides should be actively involved in product stewardship and should help to provide safe solutions to the problems of pesticide-related waste and empty containers. NGOs and farmers' organizations are invaluable channels for information and advice in communities that are poorly serviced by official bodies. They can help to implement the recommendations made in these guidelines, although they still need institutional support for their activities and cannot be expected to act entirely voluntarily. Pesticide end users also have important roles to play in minimizing the generation of pesticide-related waste and managing what waste there is in a responsible manner, in keeping with the best practices that are available.

Pesticide Management

Authorities should do all they can to develop the capacity for pesticide management and to ensure that products in the supply chain are approved and appropriate. Controls on pesticide distributors should be imposed and mechanisms developed for the management of empty pesticide containers, as well as of unused, unwanted and obsolete pesticides.

Reducing Pesticide Dependence

Pests, diseases and weeds in agriculture, public health and other situations can often be controlled without the use of synthetic chemical pesticides. Low external input methods, such as organic farming and integrated pest management (IPM) systems, have been very successful in eliminating or significantly reducing the use of pesticides in many countries and situations. Public education on the causes of pest problems and the simple hygiene that prevents them can also reduce reliance on pesticides.

Widespread implementation of low external input systems needs institutional support from government ministries, extension services, researchers and development agencies. Policies and activities that support low external input systems, including organic farming and IPM, can help to prevent the problems of pesticide-related waste as well as other health and environmental problems associated with pesticide use. National and local authorities should consider implementing policies and establishing mechanisms that support the widespread adoption of low external input production systems in agriculture and low external input pest management systems in non-agricultural situations.

Regulating Pesticide Suppliers and Distributors

Pesticide suppliers and distributors are obliged to operate within the laws of the countries in which they trade. They are also required to adhere to international regulatory controls relating to pesticides. Many international organizations such as FAO, the Organisation for Economic Cooperation and Development (OECD), the World Bank and the European Commission have guidelines or codes of practice which pesticide manufacturers and distributors should follow. Pesticide companies that belong to trade organizations also have to follow the codes of practice of those organizations.

National and local authorities can enforce their own regulations, which may include appropriate elements of codes of practice and guidelines from other organizations. Regulations should ensure that pesticide suppliers and distributors follow practices that minimize the risk of their products causing harm. Licensing systems, for example, can help enforce regulations and maintain standards throughout the pesticide supply chain.

One of the main concerns of these guidelines is that pesticide suppliers and distributors be regulated to ensure that their products are sold, stored, used and disposed of in ways that minimize the generation of waste. Suppliers and distributors should also be required, or at least encouraged, to provide ways of dealing with any waste that is generated by, for example, providing reverse supply chains to collect empty containers and unused or unwanted pesticides.

Training Pesticide Users

National and local authorities should raise awareness among pesticide users of the hazards of pesticides; the importance of avoiding them as far as possible; and the correct handling, storage and use of pesticides when their use is unavoidable. Extensive efforts have already been made in these areas. National authorities, international agencies and pesticide suppliers have implemented education and training programmes in many countries and regions. Nevertheless, in many parts of the world there is still a profound lack of understanding about the hazards that pesticides present and the ways in which these hazards can be mitigated.

People are more likely to stop generating pesticide wastes and misusing empty containers when they understand the hazards involved and how to avoid them. A wide range of media, including radio, newspapers, posters, direct verbal communication (from extension officers, for example) and formal training can be used to communicate the relevant information to pesticide end users.

Publicity and education campaigns need to be repeated, so that information is not forgotten. These campaigns should be directed towards anybody who buys pesticides, uses them and/or deals with pesticide waste and empty containers. Several different campaigns may be needed to reach all of the various groups and individuals involved. For example, in many communities women show greater concern for potential health problems than men. It may, therefore, be more effective to direct information and training to women because, even if they do not use pesticides themselves, they often pass the message on more effectively to those who do.

Information routes will also vary considerably. In some places, for instance, women rarely come into contact with agricultural extension officers, but meet health workers frequently. Or children, who may be more willing to accept warnings about their elders' uses of pesticides and may also have higher levels of literacy, can be taught in schools through reading material. Non-governmental organizations (NGOs) and farmers' organizations often work with communities that are not in regular contact with government authorities. Teachers, health workers, other community services and NGOs may, therefore, all be involved in programmes aimed at improving the health and safety profile of pesticides.

Collection Schemes

National and local authorities can help with the disposal of farmers' and householders' pesticide-related waste by establishing infrastructures for the collection and appropriate management of small quantities of pesticides and contaminated materials. In some countries widely publicized "amnesties", aimed at removing all existing waste materials, have encouraged the holders of pesticide waste to bring it to local collection points. Amnesties should be accompanied by training, education and publicity on how to prevent further accumulation of pesticide wastes. At the same time, systems that support this aim should be established. For example, arrangements can be made with pesticide distributors to take back empty containers and unused pesticides.

When collection schemes are established, the authorities should ensure that collection points and interim storage facilities for waste are secure and present no hazard to the health of people or animals or to the environment. Everybody involved in operating a collection scheme should be trained in the handling of toxic materials and have access to adequate and appropriate safety and handling equipment.

Waste materials should be transported only after they have been securely packed and only on vehicles that meet the specifications of the UN *Recommendations on the transport of dangerous goods.* Management or disposal of all the collected pesticide waste should be in keeping with the FAO guidelines on the disposal of bulk quantities of obsolete pesticides in developing countries.

Roles of Suppliers

Pesticide suppliers include the manufacturers, importers, wholesalers and retailers of pesticides. They can do a great deal to prevent the problems associated with the disposal of waste pesticides and empty containers. Everybody involved in the supply chain that delivers pesticides to end users has a responsibility to ensure that customers are well informed about, and protected from, the potential hazards of pesticides. Suppliers must, therefore, establish pesticide storage and sale practices and after-sales service that protect pesticide users, the general public and the environment.

Most countries already have legislative and regulatory systems that control the manufacture, import, distribution and use of pesticides.

Individuals or bodies who are involved in these functions as part of their commercial activities are required to adhere to national law. In many developing countries the regulatory systems controlling pesticides are less developed than they are in industrialized countries. The responsibility therefore, falls on pesticide suppliers and distributors to ensure that their products and activities comply with local and international law and best practice. Some major pesticide manufacturers implement product stewardship programmes aimed at enforcing good practice throughout the life cycle of their products. Stewardship should extend to the use and final disposal of unused products, waste and containers. Suppliers should establish mechanisms that allow pesticide end users to dispose of pesticide-related waste safely.

Storing Pesticides

When distributors store pesticides properly, containers are less likely to become damaged and the risk of chemicals leaking and contaminating other materials is greatly reduced. Good stock management helps to minimize the accumulation of obsolete pesticide stocks. Pesticide stores should be solidly constructed and secure, with good ventilation and impermeable floors that will contain chemical spills and the water used in fire-fighting. Inside the store, small containers of pesticide should be stored on shelves while larger drums and sacks are on pallets. Stores should be appropriately equipped to deal with such emergencies as fire, leakage and poisoning.

Damaged containers and contaminated materials should be safely repackaged by persons trained in the handling of toxic materials and then removed immediately from the store. Pesticides in damaged containers should never be sold for use. Similarly, pesticides that have leaked from their original containers and been repackaged should not be sold for use but should be disposed of appropriately.

Container Design

The design of pesticide containers must take into account all the specific requirements related to the safe handling of pesticides. Containers should be designed to allow safe storage, transport, preparation and use of the product, as well as rinsing and disposal of the empty container. The design of features such as the pouring orifice should allow for smooth flow of the contents and for complete emptying of the container; handles should not be integral with the container as this might create reservoirs of pesticide that are difficult to empty or rinse out.

In recent years, containers have been designed to limit operator exposure and allow efficient emptying and cleaning. However, many of them require the use of expensive and sophisticated pesticide application equipment which is not always available. In developing countries, technology, personal protective equipment and running water are often lacking and pesticide suppliers should supply products in containers that are appropriate for these conditions.

The use of soluble packaging completely eliminates the need for container disposal, but the soluble packs themselves must be contained in waterproof packaging to prevent accidental leakage of pesticides. Soluble packs must also be of an appropriately small size for farmers in the developing world to avoid the waste and other hazards that arise from the partial use of large packs. Reusable packaging that is designed for return to the original suppliers must be robust enough to withstand local conditions of transport, storage and use. Containers should be designed to allow efficient emptying and rinsing so that they are as clean as possible when returned for reuse. They should be indelibly marked so that their use and contents are easily identifiable. Lids should be permanently attached to containers to limit contamination from the small amounts of pesticide that remain inside.

Selling Pesticides in Small Quantities

Pesticide users often need only small amounts of chemicals to resolve their pest problems. Distributors should stock pesticides in original containers of a size appropriate to local needs. For example, if local farmers are likely to use only 1 litre of a product during a season, it would be inappropriate for distributors to stock and sell that product in 10-or 25-litre units.

Under no circumstances should pesticides be transferred from bulk containers to smaller ones for sale to the public. Pesticide vendors sometimes transfer products into plastic bags or bottles for sale in small quantities. The containers used may originally have been for drinks or medicines, and are generally not properly labelled. This practice is extremely dangerous because pesticide end users do not have such essential information as what hazards are associated with the product or the dose required for effective use. People may not realize that an unlabelled bottle or other container contains pesticide, and may unwittingly expose themselves to extreme hazards. The practice of decanting pesticides from their original containers into other containers is contrary to the FAO international code of conduct on the distribution and use of pesticides.

Do not encourage overbuying and stocking of pesticides. Pesticide distributors should not encourage users to buy larger quantities of pesticides than they are likely to use in a season. Sales staff often encourage bulk buying by offering lower unit prices per litre or kilogram so that the turnover of products is faster for the distributor. Pesticide end users are thereby often left with products for which they have no valid use and this may result in inappropriate use of pesticides on crops or in situations for which they are not intended; overuse of pesticides; deterioration of products that have been stored for long periods; and damage to inappropriately stored containers.

Distributors should make a commitment to take back unused products from users who have been oversupplied.

Do not sell pesticides that are inappropriate, out of date or close to their sell-by date. Farmers, householders and other pesticide end users often rely on retailers for advice about the best products for a particular purpose, and for guidance on how to use those products. It is therefore crucial that vendors are well informed about pesticides and that products are accompanied by labels and other informative material that users can understand.

Pesticide manufacturers should train retailers about which products are appropriate for particular pest problems and situations. Pesticides should only be sold for the purposes for which they are intended as indicated on the product label. For example, insecticides should not be sold for use as rodenticides, and insecticides for public health should not be used in agriculture.

Pesticides should be supplied in formulations that end users can apply effectively. For example, formulations designed for bulk application by aircraft or tractor-mounted sprayers are unlikely to be appropriate for small-scale farmers using knapsack sprayers, and wettable powders should not be sold for dry application.

Pesticide distributors and sales people should also take note of the date by which products should be used (the expiry date). When its expiry date has passed, a product should never be sold or supplied for use, even at a discounted price; its use could be dangerous. Products should not be supplied to retailers or end users within one season or six months (whichever is the shorter period) of their expiry date. The expiry date should always be clearly marked on the product package.

Take back empty containers and unused pesticides. End users of pesticides generally have neither the expertise nor the resources to dispose of empty pesticide containers and unused or unwanted pesticides safely. The burial or burning of waste are unacceptable practices because they can cause irreversible damage to human health and the environment. Pesticide distributors should help users of their products to dispose safely of empty containers and pesticides that cannot be used. Collection and disposal systems can be developed in conjunction with national or regional authorities). The following are some of the measures that provide pesticide users with safe and reliable systems for disposing of their pesticide-related wastes:

- Charging customers a deposit on pesticide containers (which would be reimbursed when empty, triple-rinsed containers were returned to the distributor) would help to discourage the market in empty containers that exists in many countries.
- Pesticide retailers should be provided with UN-approved containers (as specified in the *Recommendations on the transport of dangerous goods*) into which small quantities of obsolete and unwanted pesticides and related waste materials can be placed. Each individual waste item should be packed separately and sealed in a sturdy plastic bag to prevent the mixing of chemicals in the waste bin.
- Pesticide manufacturers and distributors should establish reverse supply chains to collect empty containers and bins of waste pesticides from retailers and transport them to sites where they can be appropriately managed.
- Information about facilities for the collection and disposal of pesticide-related waste should be well publicized to end users and should be included in training programmes, on promotional material and on product labels.
- All the waste materials and empty containers that are collected by distributors and their agents should be stored safely and securely to prevent access by unauthorized persons or by animals.
- Storage facilities should be designed to prevent materials from leaking or presenting any other risk to health or the environment.

Do not sell products that are unlabelled or labelled in a foreign language. The product label is generally the only means by which a pesticide can be clearly identified. Labels also provide crucial information on efficacy, dosage, modes and timing of use, hazards associated with the product and precautions that users should take. Pesticides without product labels should be disposed of because they may be inappropriately or dangerously used. All pesticides supplied to end users should be clearly labelled in accordance with the FAO *International code of conduct on the distribution and use of pesticides.* Labels should be in a widely understood local language and should include pictograms that clearly identify product hazards, appropriate modes of storage and handling and other precautions that users should be aware of. Products that have no label, have a damaged label or are labelled in a language other than the most commonly used local language should never be sold or supplied. This rule also applies to pesticides that have been decanted into containers other than those in which they they were supplied.

Roles of Users

Avoiding Pesticide Dependence

Pesticides are often used inappropriately in situations where simple hygiene or the forces of nature would be equally or more effective. Reliance on chemical pesticides derives from a lack of understanding of the causes of pest problems, misunderstanding of what pesticides can do, and lack of knowledge about alternative pest control methods. Training and information programmes aimed at farmers and other pesticide users should promote the principles of IPM and make as much use as possible of non-chemical measures to keep pest populations below damaging levels. Alternative measures include cultural or environmental controls, physical barriers and the encouragement or introduction of natural enemies of pests. Control measures are not always necessary, even when pests are present. A small pest population, for example, may cause only insignificant damage. When control is required, the emphasis should be on managing the pest population rather than eradicating it completely (which may, in any case, be impossible). Chemical pesticides, when used, should be carefully selected and introduced in a way that minimizes adverse effects on people and the environment. Programmes that raise awareness of the hazards of pesticides encourage users to treat pesticides with care, minimize their use, and manage them and their waste products cautiously. If pesticide users are to change the ways in which they currently buy, use and dispose of pesticides, they need to have reliable information and advice and appropriate resources available to them. Information and resources should be provided by the government and industry.

Buying Pesticides

Users should ensure that they buy only those pesticides that are appropriate for the purpose. The massive range of pesticides and commercial formulated products available makes it difficult for users to know which is the most suitable product for a particular pest problem. Before buying or using pesticides, anybody affected by an outbreak of pests should seek independent advice about alternative ways of controlling it. Advice should also be sought about the most suitable pesticide products, should they be necessary. Independent advisers include agricultural extension officers, other growers and agricultural develop-ment organizations in the area. Additional sources of useful information include newspapers and magazines, radio programmes, reference books and the Internet, when they are available. When it has been decided that a chemical pesticide must be used, distributors should advise on the most suitable product for a given task.

The product selected should not only be effective against the pest but should also be of a formulation that is appropriate to the type of application equipment used by the buyer. For example, a small-scale farmer with a knapsack sprayer should not buy a formulation intended for bulk application from crop-dusting aircraft or tractor-mounted sprayers. Products should be bought only when they are in their original, sealed containers. Containers that are damaged or appear to have been opened previously should not be bought. Nor should pesticides that have been transferred from the containers in which they were supplied by the manufacturer or importer into any other container. Pesticides are often repacked for situations and in regions where end users require only small quantities and original containers are too large. The transfer of pesticides into empty drink bottles, small plastic bags, empty cans or containers from other pesticide products is particularly dangerous since the pesticides cannot be properly identified and may be mistaken for other items such as beverages.

This has caused many cases of poisoning, particularly of children. In addition, decanted products lack the hazard information and instructions for use that appear on product labels and are, therefore, likely to be misused. Users should only buy pesticides in containers that have an original and complete product label attached. The label should be in a local language and be clearly legible. Pesticide users should always read the product label because it includes a great deal of useful and important information. Pesticide buyers or users who cannot understand the label should seek help from someone they know and trust. The label should include information about:

- the content and formulation of the product;
- its potential hazards and associated safety information;
- complete instructions for use of the product.

Labels should be designed and printed in accordance with FAO's *Guidelines on good labelling practice for pesticides.* End users of pesticides are unlikely to know about the FAO labelling guidelines, so if the label does not contain the information listed above, or is incomprehensible to the user, the product should not be purchased. Pesticides should be bought in quantities that can be used within a few weeks at most. In general, only as much product as is needed for a given pest control task should be purchased so that users do not have to store significant quantities of unused pesticides. Buyers should resist pressure from pesticide vendors to buy larger quantities than are needed, even if cost reductions are offered. They should also negotiate the return of empty containers and unused pesticides to the distributor.

The provision of free pesticides is a dangerous activity and should be prevented by regulation and good trading practices. Offers of free pesticides should be treated with caution and accepted only if the products provide an appropriate solution to an existing pest problem. Pesticides that are intended for use on major crops such as cotton or coffee are often offered for use on other crops such as vegetables, or for the control of domestic pests. Such offers should not be accepted since the use of pesticides in situations for which they are not intended can be dangerous and has been known to lead to poisoning through toxic food residues. Free pesticides should only be accepted if unused quantities can be returned.

Storing Pesticides

Owners and end users of pesticides have a responsibility to store them safely so that they do not cause harm. Badly stored pesticides are likely to deteriorate and become unusable or obsolete. The following conditions should be adhered to when pesticides are stored in homes or on farms:

- Pesticides should be kept in a secure place to which children, animals or unauthorized people do not have access.
- They should not be stored in living or sleeping quarters.
- They should be kept separately from all food, including animal feed, and away from water and water supplies.
- They should be kept dry and out of direct sunlight.
- They should be kept away from naked flames, e.g. fires or lamps.
- Storage places should be well ventilated.
- Pesticides should never be transferred into containers other than those in which they were supplied.

- Older products held in store should always be used before newly purchased products. Use-by dates on products should be strictly observed.

Empty Containers

When a container of pesticide is finished, it should be cleaned out as completely as possible. For liquid (e.g. emulsifiable concentrate) or solid (e.g. wettable powder) formulations that are diluted before application, containers should be triple-rinsed and the washings used as part of the product dilutant. Containers for dry application products should be emptied as completely as possible. Washed and emptied containers that the pesticide supplier does not intend to reuse should be punctured or otherwise rendered unusable for any other purpose. Even apparently empty containers contain pesticide residues that cannot be completely removed and they must, therefore, never be used for any purpose other than storage of the pesticides that they originally contained. In many countries, empty pesticide containers are highly valued and often sold or exchanged as storage containers for other materials such as fuel, chemicals and even food or water. Such practices are dangerous and should be prevented, for example, by puncturing any empty pesticide containers that cannot be returned to the supplier.

Empty pesticide containers should not be burned or buried. Suppliers and distributors, product labels and even national authorities often recommend these practices, but they are potentially extremely hazardous to human and animal health and the environment. Safe, hazard-free burning techniques require a good understanding of pesticide chemistry while safe pesticide burial requires knowledge of local hydrology as well as of the environmental behaviour of pesticides. Many end users of pesticides do not have such knowledge or cannot apply it to their particular circumstances. It is, therefore, strongly recommended that the burying or burning of pesticide-related waste and empty containers be discouraged rather than, as happens at present, encouraged. Whenever possible, empty pesticide containers should be returned to the distributor or taken to an approved collection scheme. If no facilities exist for the return or safe disposal of empty pesticide containers and unwanted or unusable pesticides, end users should lobby pesticide distributors, local authorities and agricultural advisers to establish schemes. The aim should be to remove potentially hazardous waste pesticides and empty containers from users and pass them on to competent authorities who have the resources to deal with them safely.

Chapter 8

Integrated Pest Management

Integrated management strategies have as their goal providing maximum economic yields to the farmer, while improving or maintaining the production site and protecting the environment. Examples of integrated pest management (IPM) procedures to improve economic yields while reducing inputs include: the integration of pesticides with cultural techniques for disease control; and biological, beha-vioural, and environmental controls of pests such as weeds. Biotech-nology promises to have a major impact on IPM through the use of recombinant DNA techniques to genetically engineer microbes for the control of diseases, insects, and weeds, and to generate plants that resist insects, pathogens, and herbicides. Examples of each management procedure follow:

1. Cultural controls: crop variety selection or host-plant resistance, double cropping and/or crop rotation patterns, minimum or reduced tillage, use of organic fertilizers, and prescription pesticide application.
2. Biological controls: natural pathogens or predators of insects and weeds. Several cultural techniques mentioned above (such as variety selection) relate directly to biological control.
3. Behavioural controls: release of sterile insects, pheromone traps, and development of lethal natural products that attract insects. These techniques have only recently received attention in IPM; they relate to pest behaviour modification based on insect communication.
4. Biotechnology: microorganisms with novel features that are currently impacting IPM strategies include viruses and bacteria

that were developed for control of insects, weeds, and/or fungal pathogens. These practices are likely to impact substantially on IPM procedures regarding the development of plants having novel characteristics such as herbicide tolerance and insect resistance.

5. Environmental controls: alterations of soil pH, moisture status, use of topography, temperature, and light intensity are among the environmental approaches. However, such manipulations are difficult to perform. While closely related to cultural practices, environmental controls relate to alterations in the physical environment to maintain yields and reduce stress.

Finally, an excellent opportunity exists to introduce expert systems and computer-aided decision-making processes into IPM strategies. Many different IPM options should be assembled into a stepwise approach that evaluates each practice, and recommends the most appropriate choice.

The best options for a specific cropping system will always depend on the probability that desired results will occur; thus, computerized applications are an obvious choice for future IPM planning and implementation.

Cultural Controls

Cultural methods can be modified by using knowledge of the ecology of cultivated plants for control of diseases, insect pests, and weeds. The five major ones are as follows.

Early Planting

Sowing crops over the winter provides a plant that is more physiologically active at an earlier period than plants sown in the spring. Accelerated crop-growth in the spring may break the plant-damaging cycle of key pests at cenoses and thus avoid the disease-susceptible stages of the crop. As an example, short season cotton has been incorporated effectively into IPM programs for cotton farming in Texas (USA) to control the boll weevil and pink bollworm. Similarly, fall sowing of sugar beets decreased damage caused by pests and diseases, including beet yellow disease and mildew of beet.

Late Planting

This technique may uncouple plantpest synchronization. For example, if wheat is sown late, it is less likely to be attacked by Hessian flies that emerge in the spring.

Crop Rotation

Sequential crop rotation may be used to reduce insect pests, plant pathogens, and weeds. Maintaining crop rotation in alternate years or at other intervals can effectively control pests. For example, planting sugar beets decreases the effects of weeds, diseases, and pests that survive in the soil and on crop debris. Crop rotation provides an inexpensive control for important weed species such as *Avena fauna* and *Elytrigia repens,* as well as for pests such as eelworms and for diseases like black leg.

Soil Cultivation

Cultivation can achieve a plentiful harvest by acting as an effective control for late infestations of weed and diseases that would lower the sowing density of emergent plants. However, if the sowing density of seeds per unit area is too low, the risk of damage from other pests increases. For example, if a sugar beet crop is planted at too low a density, then the crop can be attacked more frequently by aphids.

Reproductive Control of Pests

Technology is available to reduce the reproductive success of pests. The normal reproductive process of pests is affected more by simple, biologically effective agrotechnical operations, and less by chemical means. For example, such techniques have been used experimentally for control of *C. pomonella* and *C. molesta.*

Crop variety selection for resistance to insect and plant pathogens

For example, the technique has been employed against the Hessian fly and cabbage yellows.

Biological Controls

Biological control is agricultural management using a biological agent to reduce the impact of a specific pest or pathogen. These methods include numerous procedures such as production and introduction of insect or weed parasites, predators, and pathogens into the agreocosystems. Alternative methods include the use of pesticides that do not reduce a natural population of indigenous biological control agents and the development of more vigorous, pesticide-resistant plants, and microbes that foster plant productivity. Much recent research has focused on the use of microbes that either enhance plant development or protect plants from chemicals, pathogens, or pests. Plants and organisms used for biological control may also have the ability to cause environmental injury either through careless development or by self-replication in a susceptible environment. Research and development projects aimed at biocontrol and utilized in IPM are:

1. Mechanisms of biocontrol, such as parasitism, antibiosis, competition, stimulation of plant defences, or a combination of these factors;
2. Ecology of biocontrol agents, such as suppressive soils, physical and chemical environmental factors, composition of the soil microflora, interspecific competition, saprophytic competence, and plant stimulating or repressive effects, and
3. Formulation of biocontrol agents, such as delivery systems, carrier development, and application technology.

Behavioural Controls

Behavioural controls utilize some chemicals to modify insect pest behaviour, and control pests without the use of toxins, thereby playing an important role in area-wide control systems. At present, behavioural modification methods (e.g., pheromones) have been used to confuse or trap the male population.

This technique has been applied successfully for control of *Pestinophora gossypiella* in cotton, *Cydia pomonella* in apple orchards, and *Eupoecillia ambiquella* in vineyards. These methods are not effective with extremely large insect pest populations. Recently, chemists and entomologists have focused on relationships between plants and insects, with the aim of altering insectplant communications (i.e., antifeedants, plant attractants, and plant repellents) with chemical behaviour mediators.

A series of possibilities includes reducing the success of female insects in finding plant hosts, altering reproductive physiology, or both. Attractants have also been developed to capture female insect pests; for example, in vineyards some weed plants are more attractive to female European grape vine moths. This approach may form the basis for removing female populations that cause damage by generating offspring that directly damage commercial crops.

The visual orientation and colour preference of insects may also be used for effective control. Whitefly populations in greenhouses, cherry fruit flies in cherry orchards, and sarfly in plum orchardseach has been controlled effectively employing colour attractants.

Impact of Biotechnology on Ipm

Microbial biocontrol agents are equipped to survive and to perform the role that nature has intended. However, providing high levels of disease control in genetically uniform crops exposed to massive quantities of pathogen inoculum is not equivalent to natural evolution. Characteristics of some possible benefit to an organism in the natural state may be of little use or even detrimental when employed in agriculture. The tailoring of biocontrol agents to meet the requirements of current agricultural practices will most likely entail genetic alteration of those microbes.

Although lagging somewhat behind the advances made in the genetic manipulation of bacteria, this process has already begun in fungi. Several researchers have used mutants of biocontrol agents, either to understand mechanisms in the biocontrol process or to enhance the efficacy of a biocontrol agent. The difficulty in making mutant fungi is that mutagenesis is a relatively non-specific process: all too often, silent mutations are made in addition to the desired ones. A mutant strain may perform well when assayed for a given trait *in vitro* (such as antibiotic resistance or antibiotic production), but it may be unable to sustain itself in nature, because other undetected and later expressed mutations may make it less competitive.

Future Prospects for Biocontrol

Research on biocontrol has provided a wealth of information, not the least of which is how little we know about this phenomenon. If successful strategies for biological control of plant diseases are to be developed, much research needs to be aimed at the following questions:

1. What mechanisms are responsible for the observed effects of biocontrol agents on plant pathogens?
2. How do biotic and abiotic environments alter the viability and metabolism of biocontrol agents, and the mechanisms involved in the biocontrol process?
3. What genes are responsible for the production of biocontrol mechanisms? Can these genes be isolated, transferred, and expressed in other genomes?
4. Will genetic manipulation of biocontrol agents result in superior biocontrol strains?
5. How can biocontrol agents be mass-produced most efficiently, economically, and safely?
6. What are the most effective formulations for storage, application, and activity of biocontrol agents?
7. What are the optimum timings, frequencies of application, and treatment concentrations for biocontrol agents?

8. How readily can biocontrol agents be integrated in agriculture with other means of disease control and with crop production practices?

With definitive responses to these questions, agriculture will probably be able to make practical and economic the use of fungal and other agents to control plant diseases. The biocontrol agents alone or in combination with other means of disease control may achieve this end. Research on population ecology and genetics of parasitehost and predatorprey systems has provided new and exciting information for the control of insect pests and weeds. Authoritative responses to the following questions would probably improve substantially the value of biological pest control.

1. How can 'new association' parasites (predators) from related hosts of the pest species be selected to obtain successful biocontrol agents that will provide effective, permanent control of insect pests and weeds without further manipulations?
2. Relative to 'new associations', how important are the ecology and genetics of searching, transmission, reproductive capacity, and virulence?
3. Are there different regions (continents versus islands) where 'new association' biological control agents should be sought?
4. Since viruses, bacteria, fungi, and protozoa can be genetically engineered to improve their pathogenicity for insect and weed control, what ecological and genetic characteristics will make these microbes effective biocontrol pathogens?
5. How is it possible to genetically engineer microbes as biocontrol agents that provide effective, permanent control of a pest?

IPM clearly represents a highly promising approach that requires patient and careful application to reap high crop yields while minimizing injury to ecosystems.

Plants and Pesticides

Plants are the main recipients of pesticides, regardless of whether they themselves represent the target organism (e.g., weed) or whether the targets are pests, pathogenic fungi, etc. They are exposed to pesticides from direct application, through the uptake from soil and water, and from atmospheric drift. It has been shown that about half of pesticides applied by aircraft land outside the target cropland or forest and fall out either on adjoining ecosystems or drift into distant ecosystems.

Pesticides tend to be very reactive, mostly electrophilic, compounds that can react with various nucleophilic centres of cellular biomolecules, including DNA, or form even more reactive electrophilic products that either modify cellular components or are metabolized to more or less stable products. Ideally, pesticides should affect only the target organism; however, this ideal is rarely attained because of similarities in the basic life processes of both target and nontarget plants. Toxicity, including mutagenicity, is the consequence; it is mediated by various changes in the metabolism of plants, including formation of metabolic pathways of pesticide degradation.

Differential sensitivity of plant species to toxic and genotoxic effects of pesticides has been shown to cause overall changes in species ratios, both among weeds in a crop field and in natural plant communities, due to the reduced abundance of susceptible species with concurrent increases in naturally tolerant species. This may have further consequences for the entire ecosystem. An important factor in this situation is the occurrence of herbicide-resistant forms within susceptible plant species. Here, pesticides can play roles as agents favouring the selection of pre-existing resistant mutants and as potential inducers of genetic changes. Genotoxic pesticides are potentially able to increase mutations in DNA, controlling the expression of various qualitative and quantitative traits that could cause genetic instabilities of natural plant populations and in crop varieties.

Another important consideration is the possible formation of stable metabolites and their accumulation in plants to such an extent that they can become harmful to both human and animal populations via their respective food chains. Attention must be given to this problem despite the observation that bound residues incorporated into lignin, hemicellulose, and other carbohydrate components of the cell wall of plants are generally considered to be less toxic to the biosphere than corresponding adducts produced in animal cells.

Pesticides have also produced changes in plant metabolism and in nutritional patterns that may have secondary effects on the ecology. Even when applied at recommended dosages, some herbicides have been found to increase unwanted insects and plant pathogens. For example, when corn was treated with 2,4-dichlorophenoxyacetate (2,4-D) at the recommended dosage of 1 kg/ha, the numbers of corn leaf aphids increased by threefold; corn borers were 26 per cent more abundant and were also 33 per cent larger than the same insects on untreated corn. The larger size corn borers produced one third more eggs, and thus contributed to the overall build-up of corn borers in corn fields. The insecticides monocrotophos and phosphamidon likewise increased concentrations of nitrogen and phosphorus in rice plants; these changes were thought to contribute to a resurgence in numbers of the rice blue leafhopper.

In addition to the increased number of plant insect pests, herbicides can also increase plant pathogen attack. For instance, when corn was treated with a recommended dosage of 2,4-D (1 kg/ha), black corn smut grew fivefold larger than on untreated corn (Oka and Pimentel, 1974). Also, corn treated with 2,4-D and resistant to southern corn leaf-blight lost its resistance to the blight.

Increasing insect pest and plant pathogen attacks on crops due to the use of some herbicides may in turn increase the application of more pesticides, such as insecticides and fungicides. Thus, the environmental problem in using herbicides may be amplified beyond that of the herbicide itself. Most nutrients, especially C, K, N, P, and S, are taken up by plants, which in turn may be eaten by animals. These nutrients are eventually returned to the soil or atmosphere via decomposition of decaying organisms. The amounts and forms of nutrients in soil and plants may be changed by pesticides affecting the dynamics of these animal in the ecosystem.

Pesticides can alter the chemical composition of plants. The changes that occur appear to be specific for both the plant and the pesticides involved. For example, certain organochlorine insecticides have increased the amounts of some macro-and micro-element constituents (Al, B, Ca, Cu, Fe, K, Mg, Mn, N, P, Sr, and Zn) of corn and beans, and decreased the amounts of others. In another study, DDT, aldrin, endrin, and lindane were found to stimulate synthesis of the important amino acids arginine, histidine, leucine, lysine, proline, and tyrosine in corn, but decreased the content of tryptophane.

The herbicide simazine increased water and nitrate uptake in barley, rye, and oat seedlings, resulting in increased plant weight and total protein content. Soil treatment of carrot with insecticides Nexion, Birlane, and Dyfonate increased the concentration of free sugars in roots, whereas an opposite effect was observed after the action of the herbicide Dosanex. In this chapter attention is paid mostly to the methods and experimental approaches suitable for detection of immediate and delayed toxicity of pesticides, their harmful metabolites and residues on plants. The chapter does not describe all methods used to measure pesticide metabolites and residues, changes in the plant metabolism, or nutritional requirements caused by pesticides.

Measuring Pesticide-induced Toxicity in Plants

The toxic effects of pesticides accumulated in soil or water can be measured on intact plants grown in the natural communities or in the field, on intact plants cultivated in the greenhouse or growth chambers, or on plant cell cultures in vitro.

Plants in the field and in Natural Communities

Several endpoints are available to measure an immediate toxic effect of pesticides on crop plants in the field. Standardized procedures in field trials simulating the conditions of agricultural utilization exist, and are recommended by authorities for evaluation of toxic effects of pesticides, especially herbicides, on crops. They measure toxicity to the target organism and to crops by using endpoints such as plant emergence or seed germination, crop stand or survival, growth rate, various types of injury during plant growth (e.g., chlorosis), seed setting fresh or dry weight of plants at harvest, and crop yield. Data from these trials may serve to decide suitable doses of pesticides in fields to avoid harmful effects on crop species or varieties under specific climatic conditions.

Similar experimental protocols can be followed for studies dealing with quantitative and qualitative changes caused by herbicides among weed species in crop fields and among those in natural plant communities on roadsides, grasslands, and forests. The experimental territory is to be divided into standard plots exposed to a test herbicide or a mixture of herbicides and to untreated plots as controls. In each plot, individual species and the number of individuals within each species are determined before the application of pesticides and at various times during and after pesticide application.

Finally, plants are harvested; various endpoints are measured separately according to species and plots. Survival to harvest, dry or wet weight of plants or their parts, and seed setting or seed weight can be measured and statistically analysed using principal-component analysis or linear regression modelling.

In general, herbicides in experiments lasting even several years do not eradicate susceptible species, but rather reduce their abundance. With reductions in susceptible species, a concurrent increase in species naturally tolerant to the tested compound(s) is typical.

This situation results in an overall change in species ratios, but with no change in the number of species present. For instance, in plots sprayed annually with 2,4-D for 36 years, no new major species has become established.

Plants in Greenhouses and Growth Chambers

Greenhouse and growth-chamber conditions are especially suitable for determining the importance of individual external factors like temperature, light intensity and quality, day length, water stress, nutrition, and air humidity on the uptake, absorption, transport, translocation, metabolism, persistence, disappearance and overall toxicity of pesticides in the whole plants.

Greenhouse or growth-chamber toxicity tests on individual plant species are often used as bioassays to detect pesticides residues in soil or to assay herbicide toxicity on crop plants, but these tests can be applied to all other studies of pesticide-plant interactions. The tests are simple, rapid, and highly standardized with regard to number of plants per dose of pesticide, stage of plant development, dose range and number of repetitions, conditions of plant cultivation, positive and negative controls, and handling of data.

Principally, they follow techniques of classical growth analysis, but various protocols are available for different testing purposes. Pesticides are applied to soil or sprayed on seedlings, and their effects are measured mostly by the wet or dry weight of either the whole plant or separate shoots, roots, and leaves. Toxic effects are usually expressed by constants like LD50 or ED50 that correspond to the pesticide concentration causing 50 per cent plant death or 50 per cent reduction in plant growth or weight compared to untreated controls.

Plant Tissue Cultures

Plant tissue cultures are used increasingly in pesticide studies to screen new compounds, understand the mechanism(s) of pesticide action, bioassay metabolites, ensure experiments without microbial contamination, and save time and reduce costs. Cells in cultures rapidly absorb chemicals with which they come into contact, can be maintained for a long time, and have a high rate of metabolite formation; experiments can be standardized to a greater extent with cultures than with intact plants; and metabolic products are easily isolated and separated.

By contrast, plant tissue culture techniques in pesticide research are useless in evaluating contributions of plant-associated microorganisms or cuticular or vascular transport-mediated phytotoxicity or in testing for the phytotoxicity of herbicides acting as photosynthetic inhibitors, unless they produce additional toxic effects independent of the main pathway. Sufficient pieces of evidence are available indicating both similarities and differences in the response of plant tissue cultures and intact plants to pesticides. Consequently, in vitro test results must be validated in whole plants. Phytotoxicity of pesticides is measured either on calluses grown on agar solidified medium or on cell suspension cultures dispersed in liquid media.

The suspension cultures are more capable of being standardized and tend to produce substantial reproducibility. Various simple and standard procedures can be recommended. The toxicity of pesticides is estimated according to the depression of growth and vitality of cells or calluses. The simplest way is to measure the wet weight of calluses, packed volume of cell suspensions (determined after centrifugation in graduated tubes), or their settled volume (in nephelometric flasks). Commonly used methods include the determination of dry weight of cell suspension, measurement of cell number, optical density in sonicated cell suspension, electrical conductivity, permeability of fluorescein, reduction of triphenyltetrazolium chloride, and incorporation of 14C-leucine into proteins.

Detecting Genotoxic effects of Pesticides in Plants

Genetic changes induced by pesticides, their metabolites, and residues are expressed by various endpoints, which include:

1. Structural changes in chromosomes and chromatids, called chromosomal aberrations (breaks, deletions, inversions, gaps, translocations, rings) and other disturbances (stickiness, clumping, erosion)
2. Disturbances in the mitotic or meiotic division, like spindle inactivation, causing so-called c-mitosis, non-disjunction, and other irregularities in the chromosome distribution during anaphase, resulting in polyploid or aneuploid cells
3. Recombinational events like somatic crossing over and sister chromatid exchanges
4. Sterility and embryonic lethality
5. Mutations in somatic tissues expressed as sectors on leaves and flowers
6. Mutation in generative cells expressed either on pollen grains (in haplophase) or in sexual progeny.

To determine these endpoints, a number of plant assays have been developed, based mostly on laboratory application with test compounds, followed by cultivation of plants or their parts (roots) in the laboratory, greenhouse or field. Only a few of them are suitable or have been adapted for testing genetic effects of pesticides in situ, under conditions of their utilization in agriculture, horticulture and forestry, or to assess exposures in natural plant communities. In addition, several plant assay systems frequently used for monitoring genotoxic substance in the air and water as a part of assessing risks to humans have been employed occasionally to test pesticides. As pointed out by Grant (1982a), over 230 plant species have been used in various studies on mutagenesis; however, only a limited number of plant assays are available for routine screening for genotoxicity of pesticides.

Cytogenetic Assays

Assays for detection of chromosome aberrations in plants are some of the oldest, simplest, most reliable, and least expensive methods in the field of environmental mutagenesis. Cytogenetic abnormalities can be detected in mitotic and meiotic divisions. Analyses of somatic chromosome aberrations are carried out in cells of growing root tips, stem apexes, or in pollen tubes, whereas studies of meiosis are carried out with pollen mother cells (PMC). Most advantageous for cytogenetic analysis are the species having small numbers of morphologically distinguishable chromosomes that can be viewed microscopically by staining and preparation of squashed slides from soft meristematic tissues. These conditions are met in root tips of the broad bean Vicia faba, onion Allium cepa, hawks beard Crepis capillaris, pea Pisum sativum, in root tips and PMC of barley Hordeum vulgare, corn Zea mays, and spiderwort Tradescantia. This is the reason why these organisms have been used to test for clastogenicity of environmental mutagens, including pesticides. Techniques for staining and slide preparation are so universally known and relatively simple that many other species, including crop and wild plants, may be screened for cytogenetic changes induced by pesticides in situ or under laboratory conditions.

Two groups of cytogenetic tests are used in plants:

1. *Laboratory assays:* These are characterized by a short-term (124 h) application of test pesticides on seeds or root tips, often at higher doses than used in agriculture, horticulture, or forestry. The evaluation of chromosomal structural changes is performed during the first cell cycle after pesticide application. Mostly Vicia faba, Allium cepa, and barley root tips or seeds are used for this assay.
2. *In situ assays:* They enable application of pesticides under conditions and stage of plant development similar to that in agricultural practice. Experiments of this type can be carried

out under field conditions using plants of a single species (mostly barley, corn, Vicia faba) or multiple species forming the components of natural plant communities.

Experiments of this type have been carried out by Tomkins and Grant, who analysed 12 different species among 50 detected in experimental plots on roadsides close to agricultural fields. Plants in these plots were sprayed with selected herbicides (picloram, simazine, and diurone) at doses used in agriculture, and their cytogenetic effects were determined in somatic and meiotic cells of flower buds collected at different sampling periods. The researchers observed a significant but differential increase in the frequency of aberrant cells in various species, the highest being in the mid-season of spraying (June and July) and the lowest in the off-season.

Two other types of rapid and efficient assays in plants do not detect the cytogenetic damage itself, but rather its consequences: formation of micronuclei, and pollen abortion of sterility and lowered seed setting.

The micronucleus test in Tradescantia is based on the fact that this organism's chromosomes in early meiotic prophase are extremely prone to breakage; and, subsequently, acentric fragments lead to formation of micronuclei in the tetrade stage of pollen development. Ma (1982) has developed a sensitive assay for scoring micronuclei in flower bud cells after exposition of inflorescences to gaseous or liquid forms of environmental pollutants. The assay is very rapid, because the number of micronuclei can be determined 2430 h after the action of the tested compound. A special vegetatively propagated clone of Tradescantia paludosa, recommended for the assay, can be transferred to the exposed area either as cuttings or as plants in pots. This assay has been applied also for fumigants and several pesticides in the liquid form. Reduced pollen fertility and seed setting are assumed to be consequences of chromosome aberrations induced in the meiotic stage. Indeed, pesticides that induce chromosome aberrations in meiosis also reduce pollen viability.

The pollen sterility assay can be explored for studies in natural plant communities. This kind of research may follow the protocol described by Kurinnyj (1983), based on random sampling of flowers of various species in areas exposed to pesticides and detection of pollen viability by acetocarmine squash technique or by fluorescence techniques.

Several test systems in plants are suitable for detection of gene mutations either in many (multiple loci systems) or in one or two specific gene markers (specific locus systems). Mutations are expressed as hereditary changes in the phenotypes observed in haplophase (pollen grains), in somatic tissues of the exposed plants or in the sexual progeny of exposed plants (where mutations of a recessive character are induced in the germ cells of self-pollinated plants).

Assays for Detection of Gene Mutations

Multiple-loci system tests are very sensitive, because a variety of genetic events in a large number of loci are measured. In the most frequently used systems (i.e., in barley and Arabidopsis thaliana), based on determination of chlorophyll and/or embryonic recessive lethals, 700800 loci are involved. The multilocus system tests in plants are comparable to the Drosophila mutation assay, and are considered to be more relevant than specific-locus system tests for prediction of mutagenic response in people.

Both in barley and in Arabidopsis thaliana, test compounds are usually applied in liquid form on seeds, and the mutations are evaluated after self-pollination and formation of seeds in plants grown in the greenhouse or in the field. Whereas in barley chlorophyll-deficient mutants are recognized at seedling stage in Arabidopsis they can be detected on unripe seeds in the siliquae. The Arabidopsis assay is much less time-and space-consuming, and cheaper than the barley assay. In both systems, a large number of mutagenic compounds, including pesticides, have been screened; but seldom in situ. A similar protocol to the barley assay is used principally to detect chlorophyll-deficient mutants in wheat, rice, oats, pea, and tomato.

Much more experience in the in situ monitoring of pesticides is available with the waxy specific locus system in corn. The assay is based on the detection of mutations in the 'waxy' character of pollen grains. In brief, the starch of pollen grains having the dominant wild-type allele Wx consists of a mixture of amylose and amylopectine, and it stains bluish-black with iodine. Mutation from dominant to recessive (wx) allele leads to the loss of amylose, and the pollen grain stains red. Since over 106 pollen grains per treatment group can be scored by a rather simple technique in a short time, this genetic analysis has a statistical power comparable to that obtained in the microbial mutagenesis assays.

In the in situ experiments, Plewa et al. (1984) evaluated the genotoxic activity of pesticides in field plots (2 x 0.5 m or 10 x 3 m) situated randomly within the corn field. Insecticides and herbicides were applied in their field grade formulation at doses similar to that recommended for the control of pests and weed. The waxy system is not specific to corn only, and similar assays have been explored in barley and other species.

Mutations in haploid pollen grains can be explored for the estimation of induced changes in the gene Adh+ coding for alcohol dehydrogenase. Grains of corn having Adh-mutations stain yellow, whereas Adh+ stains blue.

Another group of specific locus mutation assays in plants, used occasionally to test the genotoxicity of pesticides, is based on specially constructed tester strains, or clones, that are heterozygotic in one or two loci. This enables the manifestation of somatic mutations shortly after the application of a test compound. Somatic mutations are expressed as sectors of differently coloured tissues on leaves (soybean, tobacco, clover, corn) or on flower petals and stamen hairs (Tradescantia). The stamen hair assay in Tradescantia interspecific hybrids is one of the most sensitive laboratory tests; it has also been used for in situ monitoring of gaseous air pollutants, giving the results 23 weeks after the exposure of flower buds.

The test has been explored also for fumigants (e.g., 1,2-dibromoethane) and pesticides in liquid form. The tobacco assay, based on the detection of green and white sectors on yellow-green leaves, seems to be very suitable for testing pesticides (e.g., maleic hydrazide) in the soil. This assay, as well as the soybean assay, detects not only reversion or forward mutations, but also somatic recombinations. A tester strain of perennial white clover, with the leaf-colour marker gene, could be a good indicator for long-term monitoring in situ, but has never been used for this purpose.

Detection of Genotoxic Metabolites Formed in Plants from Pesticides

Formation of mutagenic metabolites from xenobiotics is a well-known phenomenon in animals. Such a mechanism was not considered in plants until it was found to occur in corn sprayed with atrazine and in barley treated with sodium azide. Stable metabolites were isolated, and they caused mutations in microbial tester strains. These observations stimulated research to find suitable and efficient methods to detect mutagenic metabolites in plants formed from environmental compounds including pesticides. Recently, the state of the art in this field has been summarized by Plewa et al. (1988).

There are at least three ways to detect the metabolic activation of pesticides into mutagenic products in plants:

1. Exposure of plants or plant-cell cultures to pesticides and testing for mutagenic activity of plant extracts in any mutagenicity assay like Salmonella, yeast, or human and rodent cell assay. If extracts from pesticide-treated plants increase the mutagenicity and extracts from unexposed plants are non-mutagenic, the plant genotoxic product can be further purified and characterized.

   For atrazine's mutagenic metabolite in corn, this can be done by centrifugal fractionation, gel permeation on Sephadex, and HPCL chromatography, whereas for azidoalanin (a mutagenic product of sodium azide in barley), purification and identification are performed by other procedures.
2. Preparation of extracts from untreated plants that retain enzyme activities and their co-incubation with pesticides of interest and with an indicator organism (e.g., bacteria, yeasts, or animals cells). The activation of the pesticide to the mutagenic product mediated by plant enzymes is scored according to the frequency of mutations in the indicator organism. This system has been used successfully by numerous laboratories for pesticide studies to detect the formation of mutagenic metabolites from captan, diquat, maleic hydrazide, triallate, and ziram.
3. Co-incubation of cultured plant cells (instead of plant extract) with the pesticide and with the indicator strain for mutagenicity. This system overcomes many differences associated with the preparation of plant extracts and maintenance of their enzymatic activity.

Conclusions

This chapter deals mostly with the biological assays that are, or might be, used for detection of toxic, principally genotoxic, effects of pesticides in plants. Compared to chemical and biochemical methods, these assays are simple, inexpensive, less laborious, and able to detect toxicity without the researcher's knowing or understanding the steps that produce the ultimate effect(s).

This observation does not imply that chemical and biochemical methods are less important or less used than bioassays. Attempts have been made to detect metabolites that may be formed, and residues and impurities that may remain or that may accumulate in plants used either as human food or as animal feed and forage. In general, experimental strategies and chemical or biochemical methods of extraction, clean-up, and quantitative and qualitative determinations are the same for plants as for animals. Radioactively labelled pesticides, combined with thin-layer and high-performance liquid chromatography or gas chromatogaphy have contributed substantially to great progress in the residue and metabolite analysis in plants.

Crops and wild plants are some of the most frequent recipients of pesticides among biota. Different species and forms (varieties, cultivars) respond differently to different pesticides in terms of their uptake, translocation, degradation or activation, and accumulation of residues and metabolites, the consequence of which is differential toxicity, including genotoxicity. Nevertheless, such data on plants are generally not taken into account in the assessment of the environmental hazard of pesticides.

Chapter 9

Pesticides and Health

Issue

Pesticides, also known as pest control products, control, destroy or inhibit the activities of pests, and minimize their injurious, noxious or troublesome effects. If not properly managed, pests can affect our quality of life in many different ways. Pests can represent a threat to public health and the environment as well as create significant negative impacts to the economy if they are not efficiently controlled. The availability and use of pesticides have the potential to pose risks, both directly and indirectly, to the health, safety and well-being of people in Canada, and to the environment. As such, pesticides are highly regulated products.

Background

Pesticides have a wide range of intended uses in agriculture, forestry, industry, and society. Pesticides include conventional chemical substances as well as pheromones, organisms and devices that can be used to control a pest or to lessen the detrimental effects of a pest.

All pesticides used, sold or imported into Canada are stringently regulated by Health Canada's Pest Management Regulatory Agency (PMRA). The federal legislative authority for the regulation of pesticides in Canada is the *Pest Control Products Act* (PCPA). Before a product or use is registered, the PCPA requires the federal government to use a science-based approach to determine that the pesticide does not pose an unacceptable risk to human health and the environment. In addition, it must be demonstrated that the product has value. The use of pesticides is also subject to regulation under provincial/territorial legislation, and also by municipalities. Infants and children have been a special consideration in pesticide risk assessments for many years.

The PMRA has established policies that require additional protection for children and pregnant women, and to also take into account pesticide exposure from all sources, including food and water. These policies were formally implemented in law in June 2006 when the new PCPA came into force. The new Act also supports pesticide risk reduction. For example, it specifies that only pesticides that make a useful contribution to pest management are to be registered, and it supports expedited registration of lower-risk products.

Internationally, the PMRA continues to work with its counterparts in other countries to harmonize the processes used to regulate pesticides and ensure the protection of health and the environment. Partners include the USEPA and Mexico through the NAFTA technical working group on pesticides, the European Union, the OECD, and the United Nations through Codex and WHO.

Sources of Pesticide Exposure

Pesticide intake may occur through dietary sources as pesticides are primarily used in agriculture to protect crops and livestock. In some instances, residues of these pesticides can remain in or on the foods that have been treated. In addition, pesticide use patterns are such that contamination of drinking water sources may occur via drift, runoff or leaching through the soil. Residential exposure to pesticides may occur from uses on lawns, gardens and ornamental plants, the use of personal insect repellents, flea and tick products on pets, or uses to control pests such as ants, wasps and cockroaches in homes.

Bystander exposure to pesticides may occur from uses in institutions such as schools and hospitals, public areas such as parks and other recreational areas, and from public health uses. In occupational settings such as agriculture and forestry, exposure may occur while handling or applying pesticides. As well, workers re-entering treated areas may be exposed to pesticide residues.

Health Risks

Before a product is approved for use in Canada, it must undergo a thorough science-based risk assessment and meet strict health and environmental standards. If the use of a product poses unacceptable risks to human health or the environment, it is not registered for use in Canada. Furthermore, all pesticides registered prior to 1995 are being re-evaluated using the most modern scientific risk assessment approaches to ensure they meet current safety standards. Moving forward, pesticides will be re-evaluated on a 15 year cycle. During the evaluation or re-evaluation of a pesticide, the PMRA considers a comprehensive toxicology database in order to assess potential health effects.

These include, but are not limited to, studies to characterize acute and chronic toxicity, carcinogenic potential, reproductive and developmental toxicity, immunotoxicty, neurotoxicty, genotoxicity and endocrine disruption potential. The PMRA assessments are also informed by epidemiological evidence, general scientific knowledge and published scientific information.

The risk posed by a pesticide depends on the toxicity of the pesticide and the amount of pesticide to which a person is exposed. A pesticide with low toxicity and high exposure may cause similar risks as a pesticide with high toxicity and low exposure. Therefore, as part of its assessments, the PMRA estimates the extent to which a user and bystander(s) could be exposed through use of the pesticide. In the exposure assessment, sensitive populations and life stages are specifically addressed, including infants, children, and women of child-bearing age. Consideration is given to the different activities, dietary habits, food intake and body weight of children versus adults. A pesticide will only be registered if this estimated exposure raises no concerns. Once this is determined, the PMRA will ensure the label directions indicate the appropriate use instructions to best minimize exposure. Therefore, it is highly important that the person using the product carefully follow the label directions.

The PMRA also sets the maximum residue limits (MRLs) on food commodities, which are enforced by the Canadian Food Inspection Agency. For pesticide residues MRLs are set only after the PMRA has confirmed that any pesticide residues that could be consumed are acceptable. As required, exposure reduction measures are developed. These can include limitations on the conditions of use for a product, such as restricting use to only professional applicators where specialized equipment and training are needed.

Childrens' Health Priorities

The PMRA's human health risk assessments consider the Canadian population in general and subpopulations such as women of child-bearing age, pregnant and nursing women, infants and children. Infants and children have been a special consideration in health risk assessments for many years. Recent advances in scientific understanding reaffirm that children are not "little adults" and must be considered as a discrete subgroup. Two elements distinguish infants and children from the adult population:

1) Biological considerations. The developing fetus, infants, and children are in a state of rapid growth with cells dividing and organ systems developing. Children have a higher ratio of skin surface area to body weight than adults and on a weight for weight basis, children eat more food, drink more water, and breathe more air, than adults. As a result of these biological differences, children may absorb, metabolize, and excrete chemicals differently than adults do, potentially resulting in differing levels of susceptibility to chemical hazards.
2) Unique exposures. In addition to exposure through minute residues that may remain on some food, such as fruits and vegetables, children may be exposed to pesticide residues in breast milk and formula, through skin contact with treated surfaces while crawling and playing, and through incidental ingestion from behaviours such as hand to mouth transfer.

Aggregate and Cumulative Risk

Exposure to a pesticide may occur through different routes (oral, dermal and inhalation) and pathways (dietary, drinking water, residential uses in and around homes and schools). In order to fully assess potential risks, the PMRA conducts aggregate assessments which consider these different pathways and routes. Where it has been demonstrated that a group of pesticides share a common mechanism of toxicity, they are subject to a cumulative risk assessment in which the combined aggregate risks are assessed.

Re-evaluation of Organophosphate, Carbamate and Lawn Care Pesticides

Organophosphates Pesticides

Organophosphates (OPs) are a group of closely related pesticides used extensively in agricultural and non-agricultural sites that have potential to affect functioning of the nervous system. As a result of the common mechanism, a cumulative risk assessment of these pesticides is being conducted. Several of these pesticides have been voluntarily discontinued or have had their residential uses severely restricted. Completion of the re-evaluation of these pesticides and their uses is a priority of the PMRA.

Carbamate Pesticides

The N-methyl carbamates are a group of closely related pesticides used in homes, gardens and agriculture that have potential to affect the functioning of the nervous system. As with the OPs, a cumulative risk assessment of these compounds is being conducted and re-evaluation of these pesticides and their uses is underway. Voluntary risk reduction measures have been implemented for carbaryl, including the discontinuation of pet care products, as well as limitations on broadcast turf uses and dust formulations.

Lawn Care Pesticides

There is an increased interest on the part of the public regarding the urban use of pesticides, particularly for lawn care. PMRA initiated a special review of four herbicides and four insecticides commonly used in lawn care. The re-evaluations are completed for four of the lawn pesticides. As a result of these re-evaluations, the insecticides chlorpyrifos, diazinon and malathion (broadcast turf use), and the herbicide mecoprop are being phased out. The re-evaluation review for lawn and turf uses of 2,4-D was released February, 2005, with an update provided in August, 2006, and the proposed decision following the review of the agricultural uses is targetted for 2007. The proposed decision for the herbicide MCPA was published April 2006, and the herbicide dicamba is slated for publication in 2007. The assessment of the insecticide carbaryl is underway, and is expected to be completed in 2008.

Minimizing Your Risk

Responsible use of Pesticides

It is good practice to reduce or eliminate any unnecessary exposure to pesticides. Canadians can and should seek opportunities to minimize their exposure to and reduce their reliance on pesticides. As such, the PMRA supports integrated pest management practices, an approach combining biological, cultural, physical and chemical tools to manage pests. In doing so, pest control benefits are maximized, while health and environmental risks are minimized.

Pesticide use in and Around the Home

A homeowner must decide how much damage from pests can be tolerated before control is necessary. Before purchasing any pesticide product or hiring a pest control operator, check on possible alternatives that may be available to control the pest. Some pests may be controlled by using traps and physical barriers or by sealing cracks and crevices that may allow pests to enter the home. If you choose to use pesticides, use products only for their registered and intended uses while carefully following the directions on the label. These directions specify how you must use the product so that it poses no health or environmental concerns. To prevent accidents, pesticides should always be stored safely, in their original containers and out of children's reach. Minimize exposure to yourself and others by only using pesticides when necessary.

Food

It is always a good idea to wash fresh produce under running water and wipe dry. This will help remove any remaining surface pesticide residues, and more importantly, it will help to remove any dirt or bacteria that may be on the surface of the food. From the time when it is picked to the time it arrives in your kitchen, produce can come in contact with different environments and be handled by many people.

Using Personal Insect Repellents Safely

Parents and caregivers have always tried to protect their children from mosquito bites. Now that mosquitoes can carry the West Nile virus, there is even more concern about their bites. For most Canadians, the risk of illness from West Nile virus is low, and the risk of serious health effects is even lower. To help prevent mosquito bites, the use of a personal insect repellent should be considered. Never use personal insect repellents on children under six months of age, and for children under two years of age it is advisable to use mosquito netting around their carriages rather than personal insect repellents, unless a high risk of complications from insect bites exist. Read the label carefully before using. Pay special attention to the maximum number of applications allowed per day, the age restrictions for use, and the protection times.

Poisoning

If pesticide poisoning is suspected, consult a physician immediately. Bring the pesticide container or the label to the hospital. This may help in quickly diagnosing and treating the patient. If a pesticide poisoning or other incident occurs, including adverse effects to health and the environment, it should be reported directly to the PMRA or, starting in the spring of 2007, to the pesticide registrant. Under the new PCPA, registrants will be required by law to report pesticide incidents to the PMRA within a set time frame.

Storage

Store pesticides out of the reach of children and pets, and away from feed, food or water.

Disposal

The best solution to the problem of pesticide disposal is good planning. It is essential to purchase only the required quantity for a specific problem. Contact your municipal officials to inquire about a collection site for household hazardous waste.

Role of Governments

Federal Responsibilities

Pesticides must be registered under the PCPA before they can be imported, manufactured, sold or used in Canada. Registration under the PCPA requires a thorough scientific evaluation by the PMRA to determine if new pesticides are acceptable for a specific use and if registered pesticides remain acceptable for use once on the market.

If Canadians choose to use pesticides, they can only use a pesticide registered by the federal government for the pests and treatment areas listed on the label, and use them according to the label directions.

Provincial Responsibilities

The provinces and territories are able to enact regulations to restrict or prohibit the use of products that are registered under the PCPA in their jurisdictions. These regulations can be more restrictive than the PCPA or other federal regulations, but not less restrictive. For example, provinces and territories may be involved in the following:

- require pesticide use permits and impose additional use restrictions;
- regulate the transportation, sale, use, storage and disposal of pesticides;
- regulate the training, certification and licensing of pesticide applicators and vendors;
- respond to spills or accidents.

Municipal Responsibilities

Provincial and territorial governments may also allow cities, towns and municipalities to enact bylaws to set further regulations, including use restrictions, on pesticide use based on local considerations.

Management of Insects and Mites in Cucurbit Production

Insects serve a critical role in cucurbit crop production by pollinating flowers so the fruit develops correctly to a marketable size and shape. However, there are species of insects that can cause reductions in fruit production when abundant. These pest insects and mites must be controlled. In addition, there are beneficial insects and mites that feed on pest insects and mites and limit pest populations.

Pollination

Cucurbit crops are primarily pollinated by insects and various species of bees. The pollinating insects can and should be managed to provide effective and efficient pollination. Effective pollination results in increased yield, earliness, and quality of crops with the exception of some cucumber varieties. Male flowers, separate from female flowers, produce pollen and insects (especially bees) are essential to production by spreading the pollen. Wind does not effectively move cucurbit pollen. Therefore, pollinating insects are essential for carrying the pollen between flowers.

Factors Affecting Pollination

The following factors affect management of domesticated honey bees that are moved into production fields in hives. These factors affect other bees as well.

Weather: Cloud cover, temperatures below 60°F, and wind speeds above 15 mph reduce honey bee activity. When these weather conditions occur and persist during flowering and fruit set periods, additional bee hives may be required for adequate pollination.

Competing bloom: Honey bee hives should be located around or in fields as soon as the first fruit-producing flowers appear. A delay in introduction of bees may result in a delay in peak fruit production and reduce the crown set of fruit. Flowers on melon crops tend to be less attractive to bees than cotton and alfalfa flowers. When cotton or alfalfa grows adjacent to melons, a large percentage of bees will work cotton and alfalfa flowers at the expense of the melons. Crops that attract bees away from the target crop during the blooming period should not be planted nearby (a half-mile buffer is desirable). Similarly, many weed species are very attractive to bees and can draw bees away from the target crop. If weeds in bloom are present in or around the field, they should be destroyed before the crop begins to bloom.

Flowers: Cucurbit crop flowers are open for one day. Bees are most active in mid-morning during the summer months. They generally collect pollen during the morning and may continue activity in the afternoon but focus on collecting nectar. Male flowers appear before female flowers, produce pollen, and then fall from the plants. Female flowers that are not pollinated fall off the plant and do not produce fruit. Plants should be monitored after flowering to determine number and location of fruit set on plants.

Placement of hives: beehives should be placed in clusters around the perimeter of the field. Hives close to the field will have bees that more readily work flowers on the plants closest to the hives. Placing hives in clusters results in competition among the bees from hives and acts to push the bees further into and throughout the field.

Pesticides and irrigation: Pesticides (especially insecticides) can kill wild and domesticated bees. Therefore, pesticide applications to control insects, diseases, and weeds should be completed prior to the critical time of the crop flowering and fruit set.

If a pesticide application must be made during the flowering stage, apply the pesticide late in the evening so that flowers are not open and bees are less likely to be active in the field. Overhead irrigation may decrease pollination because the flowers fill up with water and are less attractive to bees. Overhead irrigation should be avoided during peak pollination periods.

Cucurbit Integrated Crop Management

Crop Time Line

Pre-plant	Planting	Seedling	Vining	Flowering	Fruit Set	Fruit Development	Harvest	Post Harvest
	soil insect pests—wireworms and grubs	thrips, cucumber beetles, squash bugs, and cutworms	cucumber beetles, squash bugs aphids	DO NOT APPLY INSECTI-CIDES	cucumber beetles (larvae), squash bugs, aphids, mites, melonworms and pickleworms, squash vine borers			

The Importance of Wild Bee Pollinators

By far, domesticated honey bees (*Apis mellifera*) are the most effective insects for pollinating crops. However, for crops grown in Oklahoma, wild bees can be extremely important as pollinators. Wild colonies of honey bees appear to be important, but populations have declined in recent years due to introduced parasitic mites that have destroyed honey bee populations. Other bee species are present in the state and may contribute significantly to crop pollination.

The more common types of wild bees include bumble bees, alkali bees, leafcutter bees, and carpenter bees. Many wild bees are ground nesting and some are solitary, not living in colonies as do honey bees. The major limitation of using wild bees in pollination of crops is the inability to manipulate them to focus on the target crop. In addition, the abundance of wild bees varies greatly by location and from year to year and thus they are less dependable. Honey bee colonies can be used to supplement wild bee activity.

Monitoring Pollination

During flowering, pollinating bee populations should be monitored by observing 100 individual flowers between 8 A.M. and 10 A.M. Note whether a bee is present in the flower. Because crops differ in the way pollination occurs and thus their pollination requirements, a separate discussion of the major cucurbit crops follows. Watermelon 0.25 to 2 hives per acre: Because of their flower structure, watermelons are dependent on bees for pollination. Nearly all varieties of watermelon have separate male (staminate) and female (pistillate) flowers. Watermelon pollen tends to be sticky, requiring bees to move the pollen from one flower to another. The source of the pollen, be it from a male flower on the same or a different plant, is not important for effective pollination to occur. Watermelons have a very distinct flowering pattern. Male flowers appear first, followed by female flowers 7 to 10 days later. Female flowers are open for only one day and must be pollinated on that day, or they will abort and drop off. The ratio of male-to-female flowers ranges from 4-to-1 up to 20-to-1, depending on the variety and growing conditions. Flowers open shortly after sunrise and may remain open into the afternoon. Flowers do not reopen after closing, whether or not pollination has occurred.

Each seed forms from the union of the contents of a single pollen grain and a single ovule. Successful fruit production requires the transfer of about 500 to 1,000 of the large, sticky pollen grains from a male flower to the stigma of a female flower. Although seedless watermelon does not require pollination for production of seed, it has been shown that the proper development of marketable fruit depends upon pollen stimulation of the female flower stigma. If the female flower receives an inadequate amount of pollen, the resulting melon will be lopsided or smaller on one end. For elongate varieties, a poorly pollinated melon will be smaller on the stem end of the fruit and will not contain seeds inside on the stem end. Such watermelons are usually considered culls. In addition, watermelons that are poorly pollinated are usually undersized and presence of these cull melons may inhibit the setting of additional fruit for several weeks.

Adequate numbers of honey bees must be available at the right time to provide sufficient pollination. A premium price is paid for early production, so it is best to get bees to the field within a week after the appearance of male flowers. Bees should remain in the field for approximately 1 month to ensure fruit set. In most cases, leaving bees in the field for more than a month does not substantially improve the yield but only increases the risk of pesticide contamination of the beehives. Research indicates that about 8 bee visits per flower are necessary to obtain proper fruit set and development. In order to get this level of activity, 1 honey bee colony per acre should be adequate. However, this depends on the abundance of wild bees. For dry land watermelon production, which utilizes a relatively low plant population, 1 strong hive for every 2 to 4 acres ensures good fruit set. Irrigated watermelons may require a slightly higher bee population and growers should place 1-2 hives per acre. Field surveys, in which the numbers of bees and flowers are counted along a diagonal in the field during mid-morning, are probably the best means to determine if an adequate number of bees are present. In general, a minimum of 1 bee per 100 flowers is needed to get good fruit set.

Muskmelons 1 to 2 hives per acre: Cantaloupe and honeydew melons are depend on bee pollination for good production. Nearly all information available on muskmelons comes from studies on cantaloupe. Flower structure of cantaloupes differs somewhat from watermelons, in that the “female” flowers are actually hermaphroditic (contain both male and female parts). In practice, these flowers are incapable of self-pollination, and thus pollen must be transferred from the male flowers to the “female” flowers. If no pollen is transferred to these hermaphroditic flowers, flowers shed without setting fruit. In cantaloupe, melon size is strongly related to the number of seeds present in the fruit; the more seeds, the larger the fruit. At least 1 pollen grain is required for each seed that is produced; therefore as bee activity is increased, seed production and fruit size is improved. In general, cantaloupe with fewer than 400 seeds per fruit fail to size properly and are considered culls. Increased bee activity also improves the likelihood of early fruit set (known as crown set). Crown set fruit are normally larger and higher in sugars, commanding greater market prices.

The honey bee stocking rate for muskmelons is higher than that of watermelon. In general, 1 hive per acre is needed to get good fruit set and obtain large size melons. During a midmorning survey, 1 bee per 10 “female” flowers is the minimum level of activity to maximize production. Previous research has demonstrated the economic benefit of using up to 2 hives per acre.

Cucumbers 1 hive per acre: The pollination requirements for cucumber do not differ substantially from those of muskmelon. If cucumber fruit are inadequately pollinated, deformed fruit usually result and may appear as “crooks” or “nubbins.” In addition, if cucumbers are to be machine harvested (for pickle processing), good pollination maximizes uniform production. Research indicates a honey bee stocking rate of 1 good hive per acre, but this may vary depending on variety and end use of the cucumbers (pickles or fresh-market).

Squash and Pumpkin 1 to 3 hives per acre: Squash and pumpkin have similar pollination needs because of the similar flower structure. Male and female flowers are separate; therefore movement of pollen to female flowers by insects is essential for a fruit to be produced.

Like cantaloupe, fruit size and seed set of squash and pumpkin are strongly related to bee activity. Flowers are open for one day and with squash, are usually open only in the morning. Because squash and pumpkin flowers are large, honey bees are not as efficient in pollinating them as larger, wild bees. In particular, bumble bees serve as excellent pollinators where their populations are sufficient. However, honey bees serve as the only means to increase bee activity levels where wild bee activity is inadequate. A minimum of 1 hive per acre is normally necessary although yield increases have been noted with up to 3 hives per acre.

Insect Pest Management Concepts

Recommendations for managing insect and mite pests vary depending upon the type of production. Differences in tolerance levels for pests and damage will exist for large commercial watermelon growers producing open-pollinated watermelon without irrigation, in contrast to growers that transplant seedless watermelon plants into plastic mulch with drip irrigation.

Thus, we offer a few general recommendations based on production level and additional specific recommendations for each pest. Several insect and mite species reduce plant growth and productivity when their numbers are abundant and at a time when the plants are susceptible to damage. These insect and mite populations must be monitored using common scouting techniques to determine species occurrence and population abundance. When the populations become too numerous, then the insects and mites must be controlled. Managing pest insect and mite populations has been shown to be most successful when producers follow IPM strategies.

Basic Steps for IPM Insect Management

Monitor crop growth and insect and mite pest abundance: Monitoring or scouting for insect and mite pests should be conducted in conjunction with a weekly monitoring of the fields for pollinating bees, weeds, diseases, and plant growth. Record the numbers of insects or mites per plant or leaf and average the numbers after completing the scouting of the field. Compare these numbers from week to week and among fields to determine whether pests are increasing in abundance or are greater problems in specific fields.

Early in the production season, when plants have fewer than 5 leaves or prior to vining, samples may consist of 50 plants per field. As the plants increase in size, sampling should focus on individual leaves and 100 leaves per field should be examined. Both small and large plants and old and young leaves should be included in samples each week. Scout the field by starting in a corner of the field and taking the first sample, then walk a pre-determined number of steps across the field before taking the next sample. Walking across the field in an X or Z pattern usually provides a representative idea of what is in the field. Take care to include edge as well as middle areas of the field. A hand lens or good-quality magnifying glass is a useful aid to see small pests such as aphids, thrips, or spider mites. Germinating, seedling, and vining stage plants-Check fields twice per week by examining 50 plants per field when plants are small (up to 5 leaves).

Vining stage to harvest: As plants get larger, change the sampling plan to 100 leaves per field. Take care to check the undersides of leaves, the preferred site for aphids, worms, and other insects, but also check the base of plants where squash bugs are likely to be found.

Flowering stage: In addition to checking for pests, observe 100 flowers and record number of flowers with active bees to determine whether sufficient pollinators are present.

Monitoring plant growth and fruit set : The weekly or biweekly scouting program for each field should include scouting for and recording plant growth and fruit set.

Evaluating Yield and Harvest Dates

After fruit is set, monitor the growth of the fruit during the weekly pest scouting to better project harvest dates.

Identify insect and mite pests: Use this reference manual or one indicated in the reference section to assist in identifying insects and mites found feeding on plants. When a question arises as to identification, collect the insect and/or plant part with the insect and the plant damage and place in a plastic bag, seal the bag, and place in a cool location. Take the sample to the county Cooperative Extension Service office for assistance in identification.

Decisions to take management actions: Use this manual to assist in determining whether to spend resources managing a pest. The main issue is whether the cost of the management action will result in returns greater than the cost of the control. For example, low population numbers of mites that are not increasing in number from week to week will not significantly reduce yield especially if present late in the growing season close to harvest. Therefore, the cost of application of a miticide to control the mites may simply add to the cost of production and not result in a net return to the producer.

Taking action: When deciding to control pests, choose the most cost effective method. Application of insecticides or acaricides is the most commonly chosen and most effective method for controlling pest populations when they are so numerous as to affect plant growth. Consult the Oklahoma Cooperative Extension Service for recommendations for pesticide applications for specific pest problems. Make applications as directed to ensure best control.

Evaluating controls: Within one week of applying a control measure, monitor the field to determine whether the pest population has been reduced to a level as desired. Compare the number of insects or mites per plant or leaf as recorded prior to application of the control measure with the number of insects or mites present 3 to 7 days after application of the control measure.

Insect-Mite Damage and Control

Cultural Control

Generally, plants that are weakened by poor growing conditions are more susceptible to insect pest damage and may be more attractive to pests than healthy, vigorously growing plants. Cultural methods greatly influence the extent of pest problems by 1) the selection of a good site for production, 2) the proper fertilization and irrigation, 3) the use of varieties that are less susceptible to pest attack, 4) the proper arrangement of crop plantings, 5) the consideration of crop planting dates, 6) the use of effective weed management practices in and around the cucurbit field, and 7) the timely destruction of crop residue from previous cucurbit plantings.

Biological Control

Beneficial insects and mites, diseases, and other animals feed on and kill many pest insects and mites. Many of these beneficial insects and mites occur naturally in fields. Application of broad spectrum insecticides, fungicides, and herbicides can kill many of the beneficial insects and mites and thus pesticides should only be applied if the beneficials are unable to maintain the desired level of pest control.

Plant Resistance

Many cucurbit crops have been developed to be resistant to diseases but currently there are no recommended varieties with significant resistance to insect and mite pests. However, aphids are carriers of viruses that cause devastating diseases, which cannot be controlled with pesticides. Research has shown that controlling the aphids is not effective in preventing these viral diseases. Therefore, plant virus-resistant varieties instead of trying to control aphids.

Chemical Control

Applying effective insecticides or to control insects and mites provides a cost effective method of managing pest populations. A Cooperative Extension Service educator can make specific recommendations.

Soil Insects

Insects living in the soil become significant pests when planting a crop into a field with high organic matter, especially in a newly broken pasture. Soil-inhabiting insects feed on seeds, seedlings, and roots of plants and destroy large areas of newly planted crops. Treating these pests is not routinely required; rather, the decision to treat a field before planting should be made on the basis of field cropping history and the potential of the field to have pest problems. If crops are planted in an area that was previously pasture or had a significant weed problem, apply a pre-plant insecticide to control soil insect pests. A soil insecticide works best when lightly incorporated (with a heavy chain or light cultivator) into the soil before planting. Keep the insecticide in the upper 6 inches of soil.

Seed or Root Maggots

Adults resemble small houseflies and lay eggs in the soil in early spring when soil temperatures are below 70F. Eggs hatch into larvae (maggots) and feed on seeds, germinating seeds, and developing seedlings. They are primarily a problem in soils with high organic matter; therefore, if a cover crop precedes a cucurbit crop, especially if planted in cool weather, then preventative measures should be taken.

Threshold - No treatment threshold has been developed based on scientific research. High value crops such as irrigated seedless watermelon should be treated with a soil-applied insecticide prior to planting.

Control - Plow cover crops into the soil 3 to 4 weeks prior to planting and plant after soil warms to 70F. Apply a pre-plant application of insecticide, such as carborfuran, when planting early in cool soil, and in high organic matter.

Wireworms

The wireworm adults are called click beetles and are harmless to most crops. However, the larvae are called wireworms; these are brown yellow in colour, cylindrical in shape, and grow to 1-1.5 inches long.

These larvae feed on seeds, germinating seeds, and plant roots, and when abundant, they can destroy large areas of new plantings. Monitor wireworms by placing 5 bait stations in each field 3 weeks prior to planting: dig 6-inch-deep holes and place wheat or rye seed in the bottom. Cover the hole with soil and mark with a flag. After 2 weeks, uncover the seed and check for wireworms.

Threshold - One wireworm per bait station.

Grubs

Adult grubs are called May or June beetles/bugs. Their offspring are white to off-white, with a tan head capsule, and generally wrap their bodies into a 'C' shape form. Adults lay eggs in the soil, the eggs hatch, and the larvae feed on plant roots. When abundent, they can destroy a large area of young plants.

Threshold - No threshold has been developed based on scientific research. High value crops such as irrigated seedless watermelon should be treated with a soil-applied insecticide prior to planting.

Control - Plow or cultivate fields with abundant organic matter 3 weeks prior to planting and then again at planting, or allow field to lay fallow and without vegetation for several months prior to planting. Apply a recommended soil insecticide at planting.

Above-Ground Pests

Cucumber Beetles

The spotted cucumber beetle is about $^1/_4$-inch long, yellow to greenish-yellow with 12 black spots on its back and a black head. Larvae are yellowish white with a brown head and a brown patch on top of the last body segment. They are approximately $^1/_2$-inch long when fully grown. The larvae typically feed on plant roots; however, they may also feed on the rind of melons, especially cantaloupe, and can be found under the fruit where it touches the soil.

The striped cucumber beetle is pale yellow to orange-yellow with a black head. Its wings have 3 black stripes running their entire length. There are punctures or stripling on the wings also, so that it appears as if a small punch were used to make a row of dents on the wing covers. Larvae are about $^1/_3$-inch long when fully grown and feed on roots. The larvae are white, with brownish ends and slender bodies.

The spotted and striped cucumber beetles are among the first insects to attack cucurbit crops as the plants germinate and emerge. They feed on the cotyledons and stems of seedling plants. Their chewing may result in loss of cotyledons, which weakens seedlings. Often they eat the growing point from the young plant and thus kill the plant. Also, they will chew into or half way through the stem, weakening it so that it may break off during a strong wind. Other than immediate stand loss, the greatest damage from cucumber beetle feeding is through their ability to transmit bacterial wilt. This disease does not occur regularly in Oklahoma, though when present, it can cause severe yield losses. The insect's body carries the bacterium which transmits to the plant as the beetles feed. Controlling the beetle will limit the occurrence of the disease. Beetle feeding can reduce watermelon stands. The next section, management of diseases, provides more information on bacterial wilt.

***Threshold*:** Control of cucumber beetles is critical within 2 weeks of plant emergence from the soil. In general, if 4 to 5 beetles are found in sampling 50 plants, an insecticide should be applied. Controls are generally not needed on watermelon plantings after the plants begin to flower and fruit unless beetles are very abundant. However, cantaloupe, cucumber, squash, and pumpkin varieties may be susceptible to bacterial wilt; the adult beetles must be controlled even at low numbers throughout the season so that they do not transmit the bacteria that cause the disease.

Control - Apply a recommended soil insecticide at planting. Foliar insecticides should be applied at seedling stage, as a side dress to the soil at vining stage, and/or after fruit set if the treatment threshold is exceeded or when planting varieties susceptible to bacterial wilt.

Squash Bug

Adults emerge in the spring from their over-wintering sites and search and feed on cucurbit host plants. Adults are brownish black to dark ashy black in colour and are about $^5/_8$-inch long. The body is compact and solid, with the wings overlapping on the back $^1/_3$ of the abdomen. After mating, females deposit creamy white eggs in clusters, generally on the underside of the leaves. The eggs turn metallic bronze in colour within a few hours. When the young insects (nymphs) hatch, they are pale green. They develop a grayish body colour with black legs as they mature. A squash bug passes through 5 nymphal stages, increasing in size, before becoming an adult. The entire life cycle takes 30-45 days, and there are 3 to 4 generations per year in Oklahoma. As winter approaches, egg laying decreases and the adults move out of the fields looking for places to overwinter. Nymphs present in late fall are killed by freezing temperatures and the remaining adults search for and move to overwintering sites. This overwintering population is the source of the following year infestations of adults that move into spring planted crops.

Both the nymphs and adults feed by inserting their needle-like mouthparts into the plant and sucking nutrients from the plant tissues. Overwintered adults that move into crops in early spring can cause extensive damage by feeding on small plants and seedlings. Their feeding can greatly stress and kill young seedlings. Once plants are larger, they can withstand a low number of squash bugs. Both adults and nymphs prefer feeding on leaves but will feed on all above ground plant parts.

Squash bugs are mobile and move easily among plants within a field or among fields. The insects spend most of their time within the plant canopy, mainly around the stems and on the underside of the leaves. Adults often congregate near the base of the plant and the nymphs congregate on the leaf where they hatched and then migrate to other plant parts. Squash bugs can increase in numbers very rapidly, and when present in large numbers, can cause plant wilting. When there are large numbers of squash bugs and the weather is hot and dry, the stress on plants increases significantly. Damage is noted by the sudden wilting and death of plants. Larger plants generally begin turning yellow and then wilt, just as fruit is set and maturing on the plants.

Good cultural practices can help prevent serious squash bug damage. Healthy, vigorously growing plants, tolerate feeding by squash bugs. Squash lines that exhibit some tolerance to squash bug feeding include butternut, royal acorn, and sweet cheese. Crop residues must be destroyed after harvest to reduce squash bug problems in later plantings. Following the final harvest, crop residues should be shredded and disked to reduce numbers of squash bugs that can overwinter and invade the next year's crop.

Threshold: For high value plantings apply a recommended soil insecticide at planting. For all crops treat with foliar applications of insecticide if live squash bug adults are noted laying eggs on seedling-stage plants. After plants increase in size to the flowering stage, make insecticide applications when there is an average of 1 egg mass per plant.

Control: Apply a recommended soil insecticide at planting. Foliar insecticides should be applied at seedling stage, as a side dress to the soil at vining stage, and/or after fruit set if the treatment threshold is exceeded. For specific recommendations. Insecticide applications should be timed to kill the maximum number of overwintering adults as they move into the production field. If adults successfully mate and lay eggs in the field, look for hatching of nymphs and time applications to control the majority of the nymphs. It is important to remember that small nymphs are much easier to control than large nymphs or adults. It is critical to achieve good coverage with insecticide sprays including significant penetration of the plant canopy. This usually requires a ground application of the insecticide using a tractor-mounted sprayer rather than using aerial application.

Aphids

There are several species of aphids that can feed on cucurbits, but the main one is the melon aphid. Aphids cause damage by 1) removing nutrients from plants and causing wilting, 2) transmitting virus, and 3) secreting honeydew, which inhibits photosynthesis and contaminates the fruit. Honeydew does not wash off easily and permits the growth of sooty mold on leaves and fruit.

Aphids can cause serious damage to all cucurbit crops, and lower yield even at relatively low populations. Many aphids, including the melon aphid, transmit viruses. Research has shown that the major vectors of cucurbit viruses are species that migrate through the field as they search for proper hosts. Scouts will not generally note the presence of the migrating aphids because they move into the field, land on plants, "taste" the plants by feeding, decide the plant is not a proper host, then leave the plant and often the field.

However, that single feeding by a migrating aphid is often enough to transfer a virus. The most common aphid-transmitted viruses on cucurbits grown in Oklahoma are watermelon mosaic viruses, papaya ringspot, and zucchini yellow mosaic.

These viruses stunt plant growth and cause considerable yield loss, especially in squash and pumpkin, and occasionally in watermelon. They generally appear in late summer and are worse on late-planted (fall) squash and pumpkin. Refer to the management of diseases section for more information about the viruses.

Thresholds-In general, if aphids are present on more than 10% of the leaves, an insecticide should be applied to the crop foliage. If aphid populations are near the threshold level, the presence of insect predators and parasites should be considered in making a spray decision. If the activity of beneficial insects, particularly lady beetles and parasitic wasps, appears to be increasing, sprays should be delayed for a few days to see if biological control will limit aphid populations effectively.

Control-Apply systemic insecticides to the foliage. Insecticides are rarely effective for preventing virus transmission; however, applications may limit or delay the spread of viruses within a field where the virus disease is present. The use of reflective mulches has been shown to provide significant reductions in virus occurrence. The use of mulches can be a practical, but costly, method for limiting yield losses to viruses, particularly in fall-grown cucurbits that are more likely to experience debilitating virus problems. Producers should plant virus-resistant varieties for late-season production or if virus diseases have been noted in previous years' production.

Thrips

Thrips are tiny elongate insects, usually dark in colour. Thrips feed on young plant leaves, scarring the leaf tissue, and may slow plant growth. Damage appears as small pale-yellow blotches or silver streaks on leaves, and may resemble the type of damage caused by blowing sand.

Threshold—A clearly defined threshold has not been established, but if thrip populations exceed 1-2 per plant on plants with less than 3 true leaves, an insecticide application is probably justified.

Control—Foliar application of a recommended insecticide.

Squash Vine Borer

This moth generally occurs east of Interstate 35 and is most commonly a pest of squash and pumpkin. The moth resembles a large wasp without the stinger. Females deposit eggs near the base of the plant about the time the first planting of squash begins to bloom. A small larva emerges and enters the stem of the plant. Larvae feed inside the stem or the main stalk of the plant. The feeding damage will eventually cause the stem or stalk to die, because it destroys the plant tissue that transmits food and water in the plant. When this occurs, runners or branches of a plant begin to wilt or die while the rest of the plant looks healthy. There will be sticky excrement at the entrance hole of the larva. This is generally near the base of the plant and always near the soil surface. The worm will eventually exit the stem and enter the soil to pupate. There is one generation per year in Oklahoma.

Threshold-Squash vine borers are not considered major pests of cucurbit crops.

Control-If borers have been a problem in the past, control can be achieved by using a spray schedule started when the plants begin to bloom and continued once a week until harvest is completed. Often, if a field has not had problems in the past, there is no need to control squash vine borer.

Melonworm and Pickleworm

These insect pests occur infrequently in Oklahoma. The melonworm larva feeds mainly on the foliage and in the plant terminal bud. They attack most cucurbit crops except watermelon. The melonworm is green with two white, well-separated slender stripes running the full length of the body on the upper side. The larvae are very active when disturbed and will wiggle violently. The pickleworm larva feeds on the fruit of cantaloupe, cucumber, and squash. The larval stage is green to coppery in colour, except for the brown head and a brown area just behind the head, and is about $^{3}/_{4}$-inch long. Young larvae are strikingly marked, with nearly 100 black spots evenly scattered over the body. The melonworm feeds on leaves and seldom feeds on fruit.

Pickleworm larvae can cause serious damage to cucumber, squash, and cantaloupe. Pickleworms bore into the stems and terminal buds, especially squash blossoms, early in the season. Later in the season, pickleworms bore into the fruit from the side next to the ground. They push out small masses of green, sawdust-like excrement from their holes in the fruit. After feeding for about 2 weeks, the larva moves out of the fruit to the leaves, where it rolls a leaf around its body and spends 7 to 10 days as a pupa.

Control-These insects are seldom a significant economic problem for producers. If the foliage-feeding melonworm becomes a problem, sprays used for the squash bug can easily control it. The pickleworm is more difficult to control and more damaging because it feeds inside fruit. When the pickleworm becomes a problem, a spray schedule of 5 to 7 day intervals should be followed. Use the same insecticides as used to control squash bugs. Destroy plant materials from infested fields after harvest to lessen further population build-up and future problems with pickleworm.

Spider Mites

These minute ($^1/_{60}$-inch long) creatures are not insects but are related to spiders and ticks and have 8 legs. The two-spotted spider mite is the most common. It is red and brown in colour with two spots on the back part of its oval body, though the spots cannot be seen without magnification. Mites increase in abundance during hot, dry weather and are found on the underside of the leaves. Mites reproduce very rapidly and can become very numerous in a short period of time. They lay shiny, spherical, straw-coloured eggs on the underside of the leaves. These eggs hatch in 3 days. A female lays an average of 100 eggs. Mites can complete a life cycle in 5 days when the temperature is 75°F or above. Mites spin fine silk webbing on the underside of the leaves, which is often easier to see than the mites themselves.

Mites feed by sucking out the contents from individual leaf cells. The feeding of one mite is not damaging, but mites are usually present in huge numbers. When this occurs, their feeding puts tremendous stress on the plant. Damage appears as mottling of the leaf from the many tiny feeding punctures by the mites. This area dies and the leaf turns bronze-coloured or dies. Infestations of mites often start near an unpaved road or in patches in the field and spread from these areas. Under hot, dry conditions, mite populations can build to damaging levels within a few days. To inspect for mites, use a hand lens for positive identification. Be sure the mites are alive and moving. Dead mites will adhere to the leaf and can be mistaken for live mites during inspections. Another method is to place a piece of white paper beneath leaves then rap the leaves with a pencil or finger. Wait about 15 to 30 seconds. If small dots begin to move on the paper, mites are present.

Threshold-Generally, if 10 to 25% of the leaves have mites or mite eggs, then a miticide should be applied. If the weather is hot and dry, thus favouring mite build-up, use the lower threshold.

Control-With the rapid life cycle of mites, control is based on "breaking the cycle" and stopping the build-up before large infestations become established. This requires being mindful of the weather and plant conditions. It also requires field inspection for possible build-ups. Mites often occur in patches within a field or at the edge of the field. Sometimes treatment of localized infestations can be effective, and thus could save money by treating the smaller infested area instead of treating the entire field. Effective control of mites depends on thorough distribution of the miticide to the underside of leaves and leaves within the crop canopy.

The miticide must get to the underside of the leaves where the mites are present. Do not expect one application to completely control a mite infestation. Most miticides do not kill the eggs, and those eggs will hatch within 3 days after the spray application. A hard drenching rain or overhead irrigation can control a mite infestation, especially if the rain or irrigation is followed by cloudy, cool weather. The water acts to dislodge mites and increased humidity allows the growth of parasitic fungi that kill mites. However, the increased humidity can also encourage the growth of fungi that cause plant disease, different than the fungi that kill the mites. Limit this control option if disease-causing fungi or bacteria have been a problem in this field.

Pesticide Application and Safety

Hazards to Target Plants

A pesticide applicator, employees, and customers may be exposed daily to many pesticides. Consider the pesticide's total effect on the crop and following crop(s), not just the control of the target pest. Consider such possible side effects as phytotoxicity, drift damage, and crop rotations when developing a pest management system. Producers' health and business depend on knowledge and care in application.

Phytotoxicity

The greatest hazard to the target plant is an adverse or "phytotoxic" reaction to the pesticide applied. If phytotoxicity is severe, the plant may die. Symptoms of phytotoxicity include:

- leaf drop
- stunting
- overgrowth
- discoloured foliage
- leaf curl
- stem distortion

Phytotoxicity often mimics such things as insect damage, plant disease, and response to poor growing conditions such as insufficient moisture and improper fertilization.The following items are especially relevant to the phytotoxicity problem:

- pesticide drift
- pesticide persistence beyond the intended period of pest control
- improper rate of application or improper technique
- crop rotation restrictions not followed

The cause of phytotoxicity may be easy to determine or it may be subtle and hidden.

Factors for Phytotoxicity

- High air temperature during and immediately after pesticide application Several pesticide labels warn of potential injury to plants if applied at or above certain temperatures.
- Excessive rates of pesticide application Overdosing generally does not give any better control of the target pest and can injure the plant.
- Too little water Insufficient dilution of the pesticide when mixing will result in more concentrated spray. Plants under moisture stress may be more sensitive to chemical injury.
- Uneven distribution of pesticide Inadequate mixing of pesticides in the spray tank may result in uneven distribution of the chemical. Wettable powders in particular need agitation to keep them uniformly suspended in water.
- Formulation Certain formulations may cause injury more than other types. Emulsifiable concentrates are more likely to cause injury than wettable powder. The solvents in them may allow the chemical to come into direct contact with the plant tissue by dissolving the waxy protective layer on the leaves.
- Mixing liquids or emulsifiable concentrates with wettable powders
- Variety and species differences. Some varieties or types of cucurbits may be more sensitive to certain pesticides than others. This will be stated on the label.
- Growing conditions. Weak plants in shallow soils, wet spots, or under other types of stress are more sensitive to chemical injury. Young, tender, fast growing plants with areas of new growth tend to be more susceptible to injury.

Finally, take special care to avoid injury to cucurbits when using herbicides. Some herbicides leave residues in spray tanks that injure desirable plants. Use separate sprayers for herbicides and other pesticides to eliminate this source of injury.

Drift Problems

The proximity of different plants with varying planting or harvest dates requires applicators to be especially aware of drift problems. Two types of drift are associated with pesticides. The most common is spray droplets. Spray pressure, nozzle size, wind velocity, and pesticide formulations directly affect spray droplets. Smaller droplets drift further than larger droplets. Increasing spray pressure increases the number of smaller droplets.

The other drift problem is associated with pesticides that have a low vapour pressure and is termed "vapour drift." Vapors or gases can drift in harmful concentrations, even in the absence of wind. Phenoxy herbicides and some of the newer herbicides may form vapors after application and the vapours drift to susceptible crops such as cucurbits. There are several steps that can be taken to prevent damage to non-target plants. When several pesticides are available, the applicator should strongly consider the likelihood of drift when making a choice. The applicator should use formulations and methods of application that will result in minimum drift. Pesticides should be selected that are safe for both target and non-target plants if possible. This can be accomplished by reading the pesticide label and its Material Safety Data Sheet (MSDS).

Persistence Beyond Control Period

The period of pesticide residual activity varies greatly from one class of pesticides to another. Persistence directly relates to chemical structure, rate of application, soil type or texture, temperature, soil moisture conditions, rainfall timing, and other factors. Pesticide applicators must be familiar with persistence of each pesticide that may be applied to cucurbits, especially when herbicides are used on previous crops. The applicator must also be aware of adjacent areas that may be affected.

Longevity plays an important part of pest control, since successful pest control requires knowledge of pesticide persistence to make subsequent applications. For example, herbicides used for pre-emergence weed control generally persist for 60 to 90 days, and post-emergence herbicides last from 1 to 2 days to 3 or 4 weeks, depending upon the specific mixing instructions for certain pesticide combinations and fertilizer and pesticide combinations. Many herbicide and insecticide labels list plant-back times. These dates restrict the time from application until the listed crops can be planted. Plant-back times are on labels for two major reasons. One reason is that the pesticide will damage the crops listed on the label. The second reason is the chemical company has not conducted the required studies to receive a tolerance for that pesticide on the listed crops. Thus, if the listed crop was planted, there is a likelihood of that crop having illegal residue.

Cucurbit growers should be aware that some cucurbits - especially squashes - can adsorb chlorinated hydrocarbon pesticides from the soil. The chlorinated hydrocarbons are best known as DDT, chlordane, and heptachlor. These insecticides were used extensively in the 1950s and early 1960s. Because they have long half-lives, they are still present in the soil at very low levels. Growers supplying produce to specialty markets may want to have their soil tested for the presence or absence of chlorinated hydrocarbon insecticides.

Tank Mixes

The Environmental Protection Agency (EPA) issued a Pesticide Regulation Notice addressing the question of tank mixes. This notice recognized the fact that various tank mixes had been tested and recommended by the Agricultural Experiment Stations and State Department of Agriculture, or were common agricultural practices, but directions for tank mixing did not appear on the appropriate labels. EPA has ruled that such uses are not inconsistent with the label provided the following criteria are met:

- The crop or site of application must appear on each label of the products being mixed.
- The application rate for each product must not exceed that prescribed by the label.
- The label on one or more of the products does not explicitly instruct against such mixture.

The applicator assumes the liability and risk in applying tank mixes when instructions are not specifically given on the label. Even when pesticides are approved for use in a mixture, problems may result. Formulations of a particular pesticide may vary from one product to another, and incompatibility may result from the different wetting agents, solvents, and other additives in them. Liquid formulations are particularly susceptible to incompatibility problems due to the many additives they usually contain. Incompatibility is expressed in different ways but a crystalline precipitate or a gelatinous mass is common, which would require time to clean out plugged nozzles, screens, and regulators. Another common problem is a break in the emulsion in which the different ingredients separate out, as would oil and water. This is the result of the emulsifier being destroyed, and spray injury is likely to result. Sometimes the addition of soluble salts to the spray tank will cause an emulsion to break. Separation can also take the form of particles clumping together. Incompatibility can also result in lessened activity of the combination although no visible problems can be seen. The combined materials inactivate one another, and the resulting mix will be less effective.

Hardness of water may also affect a pesticide or mixture of pesticides. Phosphate and carbonate materials are more susceptible to high alkalinity (high pH) than other pesticides. The addition of lime, lime sulfur, Bordeaux mixture, or other highly alkaline materials may cause decomposition, loss of toxicity, or phytotoxicity. *Bacillus thuringiensis* is often degraded when used with high alkaline fungicides such as Bordeaux or other copper-containing fungicides. Combinations of oil or petroleum solvents with organic chemicals are very injurious to plants. Many emulsifiable concentrates are formulated with petroleum solvents and fall into this category. Sometimes one or both pesticides precipitate or fall out of the mixture, and spray injury to the crop results. In other cases, mixing may cause excessive residues, although no precipitation is noticeable in the tank.

Most emulifiable sprays are oil-in-water emulsions producing small particle sizes. When difficulties in emulsifying an oil-or pesticide-in-water spray are encountered, correct them by adding just enough water to cover the agitator, then start the engine and run the agitator and the pump at full pressure to force the liquid through the by-pass return to the tank. If an additional or separate emulsifier is used, it should be added to the tank before adding the pesticide or oil. The pump should be allowed to run for 2 or 3 minutes. Once this emulsion is formed, add the rest of the water to fill the tank while the pump and agitator are running.

Although incompatibility may still be a problem, it can frequently be reduced to the minimum by:

- Using formulations produced by the same manufacturer
- Keeping the equipment clean and well drained
- Never putting pesticides in an empty tank The tank should always be partially filled before adding the pesticide
- Never combining concentrates, but diluting them first
- Mixing wettable powders with water to form a slurry before they are added to the tank (unless you have an inductor system)
- Adding wettable powders to the tank before the emulsifiable concentrates.

Mixture of Pesticides and Fertilizers

Many of the principles for using tank mixtures of pesticides also apply to fertilizer-pesticide mixtures. Remember these mixtures, either wet or dry, should be handled as pesticides not fertilizers. Also, unless specified on the pesticide label, water carries the pesticide. This means that to mix pesticides with fertilizer, one has to also mix the pesticide with water and fertilizer unless the pesticide label allows for mixing with fertilizer as the carrier.

Pesticides are precision tools. The total quantity of a pesticide in the tank may be less than the accepted variation in fertilizer application, which may vary from a ±10 to 15 pounds per acre. Timing, placement, and distributions are frequently different for pesticides and fertilizers. Compromising one requirement may limit the usefulness and safety of the mixture. Where tank mixtures are to be used, make sure that agitation is adequate to maintain the suspension. Suspensions also require about 1 to 2 gallons of spray carrier for each pound of product to be suspended. Some products are specifically formulated for use in fertilizer mixtures. Others specify the need to check emulsion stability and add a compatibility agent if needed.

Spray Additives

An adjuvant (supplementary) is a substance added to the spray mixture to aid or improve the performance of the main ingredient. Pesticides, whether emulsifiable concentrates or wettable powders, are formulated for general performance purposes under average conditions. For many jobs they perform satisfactorily, but there are also many situations where they fall short of the desired effect.

For example, in very hard or soft waters, a formulation may have too little or too much emulsifier, leading to mixing or excessive foaming problems. A formulation may evenly distribute a pesticide on the leaves of plants that are not waxy, but on plants with waxy leaves, the spray may form small, round droplets instead of spreading as an even film over the leaf surface. It then runs off the plant and onto the ground, leaving no deposit to protect the plant. The addition of an appropriate surfactant to the spray tank will solve this problem.

Adjuvants may be added to the spray mixture to:

- Improve the wetting of the foliage by the pesticide
- Change the evaporation rate of the spray
- Improve the ability of the spray deposit to resist weathering
- Improve the penetration, absorption, and translocation of the pesticide
- Adjust or buffer the pH of the spray solution, increasing effectiveness and longevity of the alkaline sensitive pesticides

- Improve the uniformity and amount of the deposit
- Improve the ease of mixing or compatibility of the spray mixture
- Increase the safety from spray injury to the crop
- Reduce the drift hazard to neighbouring crops
- Improve the physical properties of the mixture - for example, anti-foaming agents.

Depending upon their intended use, adjuvants are also referred to as emulsifiers, wetters, stickers, extenders, spreaders, penetrants, foaming agents, anti-foaming agents, etc.

Adjuvants are highly active materials. In most cases, a very small quantity will have great effect. Use only the amount recommended. Too much adjuvant may be just as bad as too little. It makes little difference whether the pesticide runs off the foliage because it balls up or because it forms too thin a film on the foliage. Many products contain adjuvants and additional adjuvants may cause problems such as the loss of herbicide selectivity and producing injury on normally tolerant plants. The label should be the guide on the addition of adjuvants.

Effective Spraying of Vegetable Crops

Effective control of insect, disease, and weed pests in vegetable crops depends on using proper methods of applying pesticides. The goal for applying insecticides, fungicides, and herbicides in cucurbits is to obtain uniform coverage. Uniform coverage in the plant canopy is particularly important for controlling insects and diseases. Proper sprayer design, operation, and maintenance are essential for getting the most out of pesticides.

Improving Application of Pesticides

1. Use proper boom and nozzle arrangement. The arrangement and number of nozzles positioned on a boom depends on the type of crop sprayed. Young plants can be sprayed adequately with a single nozzle per row. As plants get larger, the number of nozzles should be increased. With a sufficient number of nozzles, good distribution of the pesticide within the crop canopy can be achieved.
2. Select the correct nozzles. Nozzle selection is extremely important to get control of pests. Several types of nozzles can be used with ground sprayers, each designed for a specific use. Manufacturers of nozzles have an array of sizes and types of nozzles available. Nozzle selection should be based on 1) type

of pesticide to be used, e.g. insecticide, fungicide, or herbicide; 2) nozzle spacing; and 3) amount of pressure to be used. For insecticides, hollow-cone type nozzles are generally recommended because they deliver a smaller and more uniform droplet than a fan-type nozzle. Smaller uniform droplets provide for better penetration of the plant canopy. Alternatively, flat fan-type nozzles are best suited for herbicides where larger droplets are desirable to minimize spray drift. Recently, systems have become available to allow rapid exchange of nozzles from one type to another. These quick connections allow the change of nozzles in a matter of minutes without the use of tools. Additionally, manufacturers have recently developed nozzle technology that reduces the potential for drift from nozzles during the application of pesticides. Spray operators should look into these new improvements in sprayer technology.

3. Select proper volume and pressure. Using the correct volume of spray is essential for getting good coverage of the plant canopy. For most crops, at least 35 gallons per acre is recommended to adequately cover the foliage. Large squash or watermelon plants may require higher volumes of spray to do a good job, perhaps up to 100 gallons per acre. Maintaining the proper pressure in the spray boom is also critical for a good spray job. Lower pressures are often used for herbicide applications to reduce the potential for drift. For good penetration of insecticides and fungicides into the plant canopy, pressures greater than 40 psi are usually desirable. However, as pressure is increased, drift may become a problem. Pressure gauges should be positioned on the boom as well as on the line coming from the pump. Adjustments to the pressure should be made using the gauge on the boom as the index of final delivery pressure.
4. Check compatibility of mixtures. When using more than one pesticide at a time in a sprayer or when foliar fertilizers or other additives are used with the spray, it is important to make certain that the spray mixture is compatible both physically and chemically. Read each label to determine if any products forbid mixing with certain other products. Problems are more commonly observed when a wettable powder formulation is mixed with a liquid pesticide. If uncertain whether two products are compatible, a small-scale test ("jar test") should be conducted. To do this, add to 1 pint water, 1 $^1/_2$ teaspoon for each pounds per acre recommended of the wettable powder,

followed by 1 teaspoon for each quart per acre recommended of the liquid pesticide. Next, shake the jar and let it stand for a few minutes to see any adverse reaction of the two products. When two products are not compatible physically, the products will sometimes separate abnormally, form layers when mixed with water, or a precipitate or greasy film will form in the mixing container. If problems occur with compatibility, the two products may need to be sprayed separately, or in some cases, a compatibility agent can be added to solve the problem. When mixing different formulations, they should be added to the water carrier in the following sequence:

1) wettable powders,
2) flowables,
3) water solubles,
4) adjuvants, and
5) emulsifiable concentrates.

Pre-mix wettable powders in a separate container, then pour the slurry into the spray tank. Otherwise, wettable powders may clump and can stick to the sides of the tank.

5. Adjuvants are sometimes used with insecticides and fungicides to provide better adherence of the product to the plant. Alternately, adjuvants for use with herbicides are designed to help the herbicide penetrate the weed plant. Most insecticides and fungicides are formulated with appropriate adjuvants to improve spreading and sticking to the plant. Labels of each pesticide should be consulted to determine if a particular adjuvant is recommended or forbidden. Before widespread use, an adjuvant should be used with the pesticide on a trial basis to make sure that it works as desired and that no compatibility problem or phytotoxicity results.
6. Check the water pH. pH refers to the level of acidity or alkalinity in any material. Water varies greatly in pH depending on the source. Water used for sprays should be near pH 7.0 (neutral) and be free of suspended particulate matter. Many pesticides are sensitive to degradation at pH levels outside of the range of 6.5 to 7.5. Therefore, water should be checked with a test kit or meter and then buffered to near 7.0. Test kits or inexpensive and meters can be obtained through most agricultural pesticide dealers. Water that is above 8.0 (alkaline) can break down many insecticides in a matter of hours. When the pH of the spray

solution needs to be adjusted, the final pH (after all pesticides are added) should be checked to make sure that no alteration has occurred as a result of one of the pesticides being added.

7. `Spray when weather conditions are favourable. Sprays should be applied when spray drift and the likelihood of wash-off by rain are minimal. In general, sprays should not be applied if wind speeds exceed 10 mph or are less than 2 mph. If necessary, sprays should be applied in early morning or at night, when wind speed is typically low but not calm. If rainfall is expected within 8 hours of a spray application or if more than 2 inches of rain is likely within a day of spraying, wash-off of the spray can be a problem resulting in less than expected control of the pests. However, to control some diseases, it is better to apply the protectant fungicide on the crop in advance of rain. Be aware of the impending weather conditions and adjust the spray schedule if necessary to avoid these problems.
8. Calibrate the sprayer regularly. Because sprayer parts wear and get out of adjustment, it is very important that a re-calibration be done regularly. A slight error in the output of the sprayer can result in not enough or too much pesticide being applied. The end result is often poor pest control, wasted product, or sometimes, damage to the crop and environment. Ineffective and inefficient applications can be avoided by calibrating the sprayer on a regular basis. Several techniques can be used for calibration. One method is given in this section and sprayer operators can check the Oklahoma Extension Agents' Handbook of Insect, Plant Disease, and Weed Control, publication E-832 for detailed instructions for sprayer calibration and calculations.

Sprayer Care and Calibration

A properly maintained and adjusted sprayer is a key component for the application of pesticides. When spray equipment is worn or poorly calibrated, application effectiveness suffers and can result in wasted material and poor control of the target pest. Prior to calibrating, equipment should be serviced and all nozzles checked for uniform output and pattern and to be sure that they are all the same type of nozzle. To service the sprayer, all hoses and connections should be inspected for aging, damage, and leaks and the pump should be inspected for output and leaks. Following this, each nozzle, screen, and strainer should be cleaned and inspected, at which time the sprayer is ready for calibration and use. After the sprayer has been used, be certain to rinse out the spray tank with clean water and follow that with partially filling the tank and running the sprayer to clean nozzle tips and screens.

Selection of nozzles should be based upon the type and volume of material that will be sprayed. Nozzles come in a large array of designs and materials, but the producer should be acquainted with newer innovations in nozzle technology. Select nozzles that will spray the pesticide with the least amount of off-target drift. Generally speaking, nozzles with higher spray angles, i.e., 110° angle vs. 80° angle, can be operated at a lower boom height, thus reducing the potential for drift to occur. Other recent innovations include nozzles that create a larger droplet size, which reduces the potential for drift. There are many ways to calibrate a sprayer and most are effective.

When getting started, it may be best to calibrate using more than one method. If the same calibration is completed with two different methods, then the operator has a good indication that calibration is correct. The following steps provide one method of calibration, but other methods can be found in the Extension Agents' Handbook of Insect, Plant Disease, and Weed Control, E-832.

5 Step Calibration Process

1. To determine the distance to drive the tractor and sprayer in the field. Use the nozzle spacing on the boom, or for directed and banding sprays that utilize more than one nozzle over a given row, use the row spacing. Mark off this distance in the field.
2. Attach all equipment to the tractor that will be used for spraying and fill the spray tank half full with water. Next, select the desired gear and throttle settings and drive the course in the field that was measured off in Step 1. Be certain to provide a starting line for the course that will allow the tractor and spray rig ample time to come up to operating speed prior to starting the timing. Using a stopwatch, determine the amount of time required to drive the marked off distance. Timed drives should be repeated, recorded and those recorded times used to calculate the average time to drive the measured distance.
3. Using the average time determined in Step 2, the calibrator can then capture the sprayer discharge for that amount of time in a container graduated in ounces. For broadcast sprayers that have even nozzle spacing, capture the discharge from one nozzle. For directed sprayers that use multiple nozzles over a given row, capture the discharge from all nozzles for a given row and combine.

4. The total amount of discharge collected given in ounces from Step 3 are equal to the gallons per acre that will be applied.
5. Check each nozzle on the spray boom by repeating Steps 3 and 4. If nozzle discharge amounts vary more than 10% from the average for the entire boom, then replace the nozzle. Spray volume can be adjusted several ways including changing the speed of the spray rig, changing nozzle size, or increasing or decreasing pressure with the pressure regulator. Be aware that increasing pressure will result in smaller spray droplet size, and drift potential will likely increase. Increasing pressure to increase a spray rate should be used only when altering speed or nozzle types have already been used. Nozzle manufacturers can provide information about the flow rates and approximate pressures which a given nozzle is designed to operate. Be informed of this information and calibrate the sprayer to operate within the suggested ranges.

Disposal of Pesticide Containers

All pesticide containers should be pressure or triple rinsed immediately upon emptying. This should be done at the mixing/loading site, and the rinse water placed in the spray tank. Plastic or steel containers are to be crushed and disposed in a landfill or recycled. Bagged material may be rinsed upon emptying if the bag is lined with water-repellant material. The rinsate is placed into the spray tank for disposal. Bags can then be disposed in landfills. Neither bags nor plastic containers can be burned in Oklahoma. OSU Fact Sheet F-7462 "Rinsing and Disposing of Pesticide Containers" provides further information on various pressure rinsing devices. Never burn or bury pesticide containers.

Chapter 10

Integrated Pest Management Philosophy

The philosophy of integrated pest control is based upon:

(1) the principles of applied ecology,

(2) the recognition that 10000 mortality of all pests is not required, and

(3) avoidance of disruptive chemicals.

Since this philosophy of pest control is more of a system of principles for guidance than a set of rules in a fixed program, it requires more than the usual judgment to make it work on the farm. Integrated pest control is based on the principles of applied ecology. Whatever is done in any field situation is always founded, as far as possible, upon full knowledge of the ecology of the pests. In practice, this is attainable only by continuous sampling of the insects in all parts of the fields where integrated control programs are being applied. By this policy of watching the fields much more closely than usual, all the natural control factors can be exploited. Implementing this policy requires the services of a professional insect ecologist under contract to the farmer. This specialist in integrated pest control is the leader of a team that samples the fields, analyzing the total interacting complex of organisms, including the pests and their biological controls, and the effects of all of the agricultural practices and industrial activities of man.

This dynamic consideration of the fields takes into account the evolving patterns of plant growth, seasonal developments, and the ever–changing populations of organisms associated with them. Emphasis is always on the developing insect population levels and the predictions as to the course that the pest populations will take.

The supervising insect ecologist implements successful pest management in the fields by applying any or all practical cultural and biological practices known to suppress the pests. To attain a favourable balance in a field, it is necessary to attract, augment, and conserve all possible numbers of beneficial predators, parasites, and disease organisms in the fields. Even low numbers of insectary-grown and/or field harvested beneficial insects can be colonized.

Best results from integrated pest control are obtained when effective natural enemies of most target pests are present in the complex of organisms. Crop rotations to alfalfa, grains, and pasture, along with planned cover crops, bring greater plant diversity to a farming area. This diversity of plants brings greater diversity in both host and prey species, and provides nectar, pollen, and alternate host organisms as food for predators and parasites.

Alfalfa, grains, and pasture grasses are normally under such satisfactory biological control that they usually require very little control other than cultural management. Such management programs in alfalfa, using strip harvesting techniques, aids in preventing sudden emigration of pests at cutting time. With less disruptive action to pests and their biological controls, there is greater stability in population interaction, and less chance for one species to dominate and cause damage. Different natural enemies have different efficiency levels related to pest population densities; the above environmental manipulations help provide a complete complex of predators and parasites, at a very low cost.

Field harvesting techniques can selectively remove portions of pest populations and collect many predators, parasites, and disease organisms for use in augmenting the natural controls in other fields. Releases of this material, along with colonizations of insectary-grown beneficial insects, can help bring the field into a favourable balance in time to prevent economic losses. Every pest spot in a field that is controlled biologically adds to the total number of predators in the field; every biologically controlled field grows more predators for the nearby fields and the agricultural area. The less you spray, the less you need to spray. This complex of beneficial insects is a natural resource that the farmer must protect and conserve by applying pesticides only when absolutely necessary to protect the crop. For certain pests there are no immediate prospects for development of effective pest suppression techniques. Then pest populations increase beyond tolerable limits and there is no predictable chance of obtaining a favourable balance, the crop can always be sprayed.

This leads us to the second principle of integrated pest control; a recognition that 100% mortality of all pests is not required to prevent economic losses. This may seem elementary, but society has been brainwashed into believing that "the only good bug is a dead bug"; it is important to point out that insects, mites, weeds, etc. are pests only when they affect our way of life beyond tolerable limits, It is essential to the concept of integrated control that we live with a few pests in order to obtain and retain a favourable balance.

Strict standardization procedures that do not necessarily give the consumer food of better quality or taste may force the farmer to spray for light infestations of a pest. Often any plant damage that might occur would not be harmful to the crop, except perhaps in appearance. Plants can tolerate low populations of aphids, thrips, and mites with no intolerable damage affecting profitable production of the crop. Instead of spraying, the insect ecologist uses these minor pests as food in colonizing additional insectary-grown or field harvested predators.

The challenge for the future is to devise ways to grow and or field harvest sufficient beneficial insects and apply them to the fields by mechanical means; in this way, biological control could be better implemented and guaranteed, and many more fields could be saved from pesticides.

The third principle of integrated pest control is the avoidance of disruptive chemicals. Not all pesticides have adverse effects, and certain dosages of conventional pesticides are less disruptive to the existing biological control. Backup sprays are applied when necessary, but of a kind and in a manner least disruptive. Integrated pest control seeks to suppress target pest populations below tolerable damage levels without destroying their biological controls.

The professional insect ecologist must know how to use pesticides in this way in given situations. However, because of the polluting side effects of too much pesticide in the environment, it is even more important that he know when and where not to use them.

There is not time to dwell on all the pest control problems that have developed as a result of complete reliance on chemicals; a short list would include target pest resistance to pesticides, target pest resurgence following treatments, secondary pest outbreaks, pesticide residue problems in the crop, the soil, or drift to nearby areas, or runoff to streams and drainage, and destruction of pollinators and other forms of life beneficial to man.

In many instances these side effects have created havoc, and the undesirable results have more than off-set any original benefits. It is becoming more and more difficult to kill the target pests economically.

One often-underestimated side problem is the drift of pesticide residues onto nearby crop areas and the widespread effect of destroying favourable biological balances. Searching predators and parasites moving over the plants in search of prey are exposed to more residue, and are killed in greater numbers than the more sedentary pests. This uncontrollable drift creates new pest situations in nearby fields, that in turn require treatment: the problem is spread further and further. Such drift effects have triggered pest outbreaks over large areas and destroyed many beneficial insects. The more you spray the more you need to spray.

The cost of chemical pest control programs is high when all of these side effects begin to interact. The crops may be saved, but at great cost to society and the farmer (who is stuck with the pesticide bills). The real measure of success in farming is in the profits; the real problem that causes the farmer to look for new answers to pest control is the spiraling cost of repeated and ineffective treatments. This search for new answers has forced farmers and entomologists to modify their pest control practices, so as to avoid disruptive chemicals as much as possible.

How effective is integrated pest control in practice? Experience has shown that this philosophy, when practiced on the farm by competent professional insect ecologist, is the only pest management system dynamic enough to solve the problems and keep up with the changing agricultural problems of the future. The farmer must willingly support a team effort to carry out practical application of integrated pest control. The leader is the insect ecologist, a highly trained specialist who utilises materials and techniques as complicated as those used by the pharmacist or veterinarian. New, more sophisticated integrated pest control programs of the future will be implemented by these professional insect ecologists.

The backup part of the integrated pest control team is the insectary, which grows the predators and parasites, and field harvests other beneficial insects, for use in applying positive biological and environmental pest suppression. Other members of the team are the researchers, specialising in long-term ecological studies and development of integrated pest control techniques. Laboratory research specialists on physiology and behaviour can discover much to aid in interpretation of data, but only studies in the fields will reveal the true interplay of all the complex ecological factors. The usual fragmentary short-term experiments, so common in pesticide evaluations, are more than vulnerable to error, when considered in the long-term ecological sense.

It is important to understand that valid scientific discovery can result from careful observations and recordings of recurring natural happenings. In other words, scientific proof can be obtained from a sequence of experiments or experiences. History will repeat itself, provided the same combination of effective conditions recurs. In applying integrated pest control, the insect ecologist frequently samples and assesses the fields in his care, constituting a continuous experiment in applied ecology not duplicated at research level. This experiment is carried out in many different field situations, throughout the season, and often over many years on the same farm. What better proof is there of the success of integrated pest control than the practical experience of profitable production of quality products?

Research specialists in integrated pest control must work closely with the insect ecologist, who is the foreman in charge of pest control. Since a multifactored environment is not easy to understand, the researcher can aid in interpreting the experience of the applied worker. At the same time, the insect ecologist can provide assistance to the researchers economically, gathering samples, etc., because he lives in the area of the field experiment. In this way the practical limitations of time and money can be somewhat overcome, and the work of solving the pressing problem of pest control can continue. The commercial pest control applicators and the producers of chemicals and other pest control products will continue to make an important contribution. The pesticide companies will make new products to answer the needs of integrated pest control, because it is the more selective target pesticides that are least disruptive that the farmers will buy.

Active and aggressive educational programs supporting the ecological approach to pest control can help separate fact from fiction regarding integrated pest management. General recommendations to farmers should be flexible, to allow time for alternative integrated pest control techniques to be exploited. Adjustable economic injury levels, including the assessment of beneficial organisms and suggestions for use of selective pesticides, are advisable.

The combined efforts of all members of the integrated pest control team are needed to insure successful and profitable production of sufficient food and fibre for the future. Successful pest control is relative, but in the final analysis it is measured in the economic sense. High costs of ineffective pesticide programs and their ultimate effects on food and fibre prices concerns farmer and consumer alike. There is no question of the success of integrated control in the minds of farmers who have changed their philosophy of pest control and are "suddenly" not paying excessive pesticide bills. The first year or two of successful integrated pest control is often attributed to "luck" or a "good year" by critics of such programs.

After several years of economical pest management with little or no spray bills, the farmer eagerly support their insect ecologist and the insectary that supports him. As food supplies become shorter in the coming decades, and as consumers become more conscious of high food costs and toxic residues in foods and less conscious of appearance only, the artificial standards that pesticides depend on will be relaxed. More sophisticated and safer processing of farm products will be developed to insure removal of insect parts, with the result that the consumer will have bug-free as well as toxicant-free products.

Insect Management

A successful insect management program can best be accomplished by combining IPM techniques, such as accurate pest identification, scouting, monitoring and action thresholds, with biological and alternative (preventative) control practices and selective insecticide applications.

Use scouting and monitoring techniques to help assess and quantify insect populations over time.

Insecticides should be used only when action thresholds have been exceeded or damaging insect populations are present. Action thresholds help minimize insecticide use and crop damage. Avoid making routine insecticide applications to crops without evidence of insect activity and damage. All insects have natural enemies which, if conserved, can help regulate pest populations.

Whenever possible, use selective insecticides that spare beneficial organisms and target the pest you wish to control. Broad-spectrum insecticides should be used as a last resort. All insecticide applications should be made with ample water and with nozzles directed so that they provide thorough coverage of the plant parts where insects hide. Alternate between insecticide classes or families to help manage insect resistance and extend the life of available products.

Insect Identification

Become familiar with the biology and life-cycle of the major insect pests that attack crops on an annual basis. Understanding some basic insect biology often reveals when the pest is most vulnerable to control measures and helps lead to successful management efforts. Insects usually have either a simple life-cycle, where they grow from egg to nymph to adult, or a slightly more complex cycle, where they mature from egg to larva and then go through metamorphosis in the pupal stage before becoming adults. Insects damage plants either as nymphs/ larvae or as adults, or in both immature and adult stages.

Learn to recognise the important life-stages of major insect pests, and to recognise the crop injury or damage they produce to help determine when control efforts are needed. You should also be able to identify common beneficial insects and their immature stages (i.e., lady bug larvae) to avoid accidentally targeting natural enemies with insecticides. There are dozens of minor and secondary insect pests that may attack crops on a less frequent basis. Growers should have unknown insects and minor pests identified when they are suspected of causing crop damage. Misidentification often leads to the application of an ineffective pesticide and extensive or chronic crop damage. The application of ineffective or unnecessary pesticides can often reduce the populations of beneficial organisms or natural enemies and lead to secondary insect or mite outbreaks. An accurate diagnosis of the problem early on can often avoid years of frustration and needless expense.

Scouting and Monitoring

Crops should be inspected or scouted for insects or damage in a systematic fashion, on a regular basis throughout the growing season. For many crops and pests, this may mean walking fields weekly, or even more frequently, especially during critical or vulnerable plant development stages. Crops should be scouted in a systematic fashion by walking in an "M" or "W" pattern as you crisscross the field. Select plants (i.e., 25 or 50) at random and quantify the pest damage or count the individual insects. Record the average number of insects or damage per plant for each field. Scouting crops always saves money in the long run by allowing for early pest detection, by reducing crop damage and by helping to maintain consistent quality. Sometimes scouting duties can be performed by other farm members or by hired consultants.

Monitoring insect populations with various types of traps (black light, pheromone, sticky, baited) can supplement or sometimes substitute for information normally gathered during crop scouting. Insect traps can help you quantify pest pressure that is difficult to see, such as the number of night flying (i.e., corn earworm) moths present. The number of insects captured in traps is often used to time scouting activities, predict future pest levels, or is used in conjunction with action thresholds to time sprays and help avoid crop damage.

Action Thresholds

Action thresholds tell you when to spray to prevent economic damage to the crop. They also help you avoid applying insecticides to crops when insect populations are low or no pests are present. Thresholds can be based on the number of insects found per plant, the amount of injury or damage per plant, or the number of insects captured in a trap. Using action thresholds helps improve insecticide timing and effectiveness, helps minimize the number of applications and associated costs, and helps reduce crop damage. Use action thresholds whenever possible to help you decide if and when a spray is needed. Some insect thresholds are provided in this publication.

Preventative Controls

As mentioned in the introduction to the general Pest Management section, there are many preventative management options, such as cultural, mechanical, physical, genetic and biological controls. These should be used whenever possible to help prevent insect pest problems. Preventative insect control options can be used alone or in combination to provide a complete management program. For example, plant inspections, eliminating weeds in the greenhouse, using screens, avoiding excess nitrogen applications, using plastic mulch and preserving natural enemies can combine to produce a very effective aphid management program. If the distance between fields is great enough, crop rotation (alone) for Colorado potato beetle and cucumber beetles can often keep these tough pests from reaching damaging levels for the entire season. However, if only nearby fields are available, CPB can be effectively controlled by the use of a combination of two or more alternative controls, such as short-distance crop rotation, intervening trap crops or plastic-lined trenches, row covers, organic mulches, flaming, and microbial controls. Many preventative controls are mentioned in crop-specific IPM manuals, individual pest fact sheets and on IPM web sites.

Perimeter Trap Cropping utilises a combination of control measures to concentrate and/or kill pests in the border area of the field. This technique involves planting one or more rows of an attractive plant species so that it completely encircles the less attractive main crop, and intercepts a migrating insect population.

Often a chemical or microbial insecticide is used to spot-spray the perimeter trap crop (only), sparing beneficials on the main crop, and substantially reducing insecticide use. Perimeter trap cropping has been shown to be effective at controlling diamondback moth larvae on cabbage, using collards as the trap crop; pepper maggots on bell peppers or eggplant, using hot cherry peppers as the trap crop; and cucumber beetles on summer squash, winter squash, cucumbers, melons, and sometimes pumpkins, using Blue Hubbard squash or another Cucurbita maxima variety as the trap crop. This technique may prove useful on related crops and against additional pests in the future.

Insecticides

Follow the label specifications for application rates. If the rate suggested in this guide does not agree with the current label, follow the label recommendations. Amounts of pesticides are in lb/A (pounds per acre), oz/A (ounces per acre) or pt/A (pints per acre) of commercial formulation, not a.i. (active ingredient), unless otherwise stated. The percentage by weight or concentration of a.i. varies with different brands of pesticides. The label gives pounds of a.i. in each commercial preparation. For example, Lorsban 4E has 4 lb active ingredient/gallon of formulated material.

The amount of insecticide recommended per acre should not be changed when varying gallonage of water per acre. Make sure a pesticide is labelled for concentrate application before using a low volume sprayer, air or mist sprayer, aircraft or other concentrate application equipment.

Seed Treatments

Insecticide seed treatments may help protect your crop from soil insects and other early-season pests. Some of the seed treatments listed under various crops must be applied by a professional licensed seed coating applicator. In some cases the use or purchase of treated seed may be economical as it may result in a substantial reduction in pesticide use and increase the plant stand and yield in the field.

Resistance Management

Adult pests that survive an environmental hardship, such as the application of an insecticide, are likely to pass the trait that enabled them to survive on to their offspring. Repeat applications with the same type of pesticide will eventually remove almost all the susceptible individuals from a pest population and leave only those with the resistant gene. Because insects and mites go through generations quickly, resistance to an overused pesticide can develop in as little as three years.

To preserve the useful life of a pesticide, it is necessary to practice resistance management. The most effective way to extend the useful life of an effective product is to use it on a single pest generation only, and then on the next generation use a second pesticide with a different method of killing the pest (mode of action). On some pests particularly prone to developing resistance (i.e., Colorado potato beetle) it is best to use a product with a particular mode of action on one generation every second year only. To help select pesticides with a different mode of action.

Resistance Groups (number and letter) indicate products with a common mode of action. For multiple applications to the same crop, select products from different resistance groups. Once a pest develops resistance to a group of pesticides with a particular mode of action, a higher rate of a similar chemical from the same group usually will not control the insect.

There are many other techniques that can help delay the onset of resistance. Most resistance management techniques help minimize the use of pesticides, so that a lower proportion of each pest generation is exposed to the toxin.

- Integrate chemical control with effective cultural, mechanical, physical, and genetic management options.
- Use biological/microbial control agents or other selective pesticides to preserve natural enemies and help minimize the number of repeat applications.
- Scout, monitor and use action thresholds to ensure that applications are necessary.
- Good spray coverage helps do the job right the first time and avoids unnecessary repeat applications: use the proper size nozzles and the correct angle or orientation and an adequate amount of water per acre.
- Time the application so that the most vulnerable insect life stage is exposed to the spray.
- Use spot sprays, perimeter trap crop treatments, refuge plantings, and other methods that prevent the entire field or population from being treated to help preserve susceptible individuals.

Toxicity Rating of Insecticides

Insecticides vary greatly in their toxicity to humans and the environment. The toxicity of the insecticide is usually stated in the precaution on the label. For example, a skull and crossbones figure and the signal word "Danger" are always found on the label of highly toxic (Toxicity Class I) materials. Those of medium toxicity (Toxicity Class II) carry the signal word "Warning." The least toxic materials (Toxicity Class III) have the signal word "Caution." The toxicity of a pesticide is expressed in terms of oral and dermal LD50. LD50 is the dosage of poison that kills 50% of test animals (usually rats or rabbits) with a single application of the pure pesticide for a given weight of the animal (mg/kg of body weight). The lower the LD_{50} value, the more toxic the material. Oral LD50 is the measure of the toxicity of pure pesticide when administered internally to test animals. Dermal LD50 is the measure of the toxicity of pure pesticide applied to the skin of test animals. Generally, an oral exposure is more toxic than a dermal one.

Vegetable Pest Identification

In nature, there are always some garden pests chewing on plants; that's just the way it is. However, not all pest damage is significant enough to warrant action. Even the healthiest gardens encounter bugs at one time or another, yet they still produce a beautiful harvest. As gardeners, we must each consider the level of pest activity that we are willing to tolerate. Every garden is susceptible to a pest invasion - tucking into foliage, feasting on the vegetable patch and burrowing through the bedding.

The best way to maintain a healthy garden is to educate yourself and learn to identify common "bad bugs." Inspect your garden regularly to detect problems early. The sooner a pest is identified the easier it will be to manage using earth-friendly methods.

Ants

Ants are a common pest in the garden, especially on the lawn, sandy ground, plants and fruit trees. They are prolific when it comes to creating nests and usually do so in a sunny area of the garden. They actually do little serious damage to plants, although they can be a nuisance and weaken both garden and greenhouse plants.

How to Treat Ants

Have a dust around plants and along runs with a good ant killer powder or spray, which are widely available. For chemical-free ant control, homemade pest repellents provide the best form of prevention, and there are certain smells they detest. Try out mint, camphor and clove oil. You can also grow your own ant repellent in the vegetable garden, by mixing hot peppers in a blender with a bit of water to create a thick mash, which you can spread in problem areas. Use a stiff brush to remove ant hills in the lawn.

Black and Greenfly (Aphids)

Greenfly or blackfly - also known as aphids - are a major plant pest, sucking the sap of many plants. These small green or brown insects gather on shoot tips, flower buds and the undersides of young leaves, and are very common on all types of plants. They can also deposit a sticky honeydew on which disease spores can stick, and leaves may twist or yellow.

How to Treat Aphids

Tackle aphids as soon as they are spotted. Spray with a greenfly killer insecticide or, for an organic option, as aphids breathe through their skin, spraying with a diluted washing up liquid will clog their skin and cause them to suffocate quickly. For plants which are affected outdoors, try to encourage birds with seeds and tables near plants at risk from aphids as they will feast away.

Bulb Fly (Daffodil & Narcissus Bulb Fly)

This small black and brown fly swoops just as the foliage is dying down on the bulbs, after flowering, and lays eggs in the crown of the bulb. The eggs hatch and become the maggot, which bores down into the bulb to feed on the inside. Daffodil and narcissus bulbs become soft, squidgy and rotten when the inside is eaten out by the fly's maggot, and very few leaves or flowers will be produced.

How to Treat Bulb Fly

Eliminate bulb flies with a strong mixture of insecticides and water, or for organic control, apply diatomaceous earth, available at garden centres, to help control freshly hatched grubs. Consider covering over the planting after bloom with a lightweight garden cover, such as fine netting.

Planthoppers

A planthopper is any insect in the infraorder Fulgoromorpha within the Hemiptera. The name comes from their remarkable resemblance to leaves and other plants of their environment and from the fact that they often "hop" for quick transportation in a similar way to that of grasshoppers. However, planthoppers generally walk very slowly so as not to attract attention. Distributed worldwide, all members of this group are plant-feeders, though surprisingly few are considered pests. The infraorder contains only a single superfamily, Fulgoroidea. Fulgoroids are most reliably distinguished from the other members of the classical "Homoptera" by two features; the bifurcate ("Y"-shaped) anal vein in the forewing, and the thickened, three-segmented antennae, with a generally round or egg-shaped second segment (pedicel) that bears a fine filamentous arista.

Nymphs of many Fulgoroids produce wax from special glands on the abdominal terga and other parts of the body. These are hydrophobic and help conceal the insects. Adult females of many families also produce wax which may be used to protect eggs. Planthoppers are often vectors for plant diseases, especially phytoplasmas which live in the phloem of plants and can be transmitted by planthoppers when feeding.

Cixiidae

The Cixiidae are a family of fulgoroid insects, one of many families commonly known as planthoppers, distributed worldwide and comprising more than 2,000 species from over 150 genera. The genera are placed into three subfamilies, Borystheninae, Bothriocerinae and Cixiinae with sixteen tribes currently accepted in Cixiinae.

Cixiid species are typically comparatively small (body size less than a centimetre) and usually inconspicuous. Nymphs live underground, feeding on roots. Adults feed on herbs, shrubs and/or trees; some are polyphagous, while others are specialised on their host plants (monophagous).

A couple of species are cavernicolous, feeding on roots in volcanic caves. Females occasionally bear impressive "wax tails" produced by wax-producing plates at the tip of their abdomen.

Several species are of economic importance (e.g. *Hyalesthes obsoletus*, *Haplaxius crudus*). Phytoplasma are common parasites in these insects, causing diseases in coconut palms and foliage, grapevines, sugar beets, and lilies.

The fossil record of Cixiidae is limited, and a number of taxa which have placed into the family may need to be reexamined and moved to different families. The oldest confirmed taxa are from the Early Cretaceous with *'Cixius' petrinus* described from Barremian deposits in England, *Karebodopoides aptianus* from Hauterivian to Aptian Lebanese amber and Cretofennahia cretacea plus an unnamed specimen from the Aptian of Brazil.

Due to the abundant nature of Cixiidae as inclusions in Eocene Baltic amber a number of taxa have been described, including *Glisachaemus jonasdamzeni* and *Autrimpus sambiorum*. Several taxa have also been described from Miocene Dominican amber including *Oligocixia electrina* and *Oliarius kulickae*.

Many gardeners are misidentifying the white insects on plant stems. In most cases it is planthoppers rather than mealybugs. The planthoppers are not destructive and jump away when touched. Mealybugs stay put. A water blast is all that's needed for the planthoppers. The insect problems are interesting and diverse this season. For mealybugs, I would probably try the EcoSMART insect control product.

The line is in quite a few of the retail stores now. Another option is to spray the Mound Drench mixture. It has worked for me in the past. Mix equal parts orange oil, compost tea and liquid molasses. Use 4 oz per gallon of water for the spray formula. It can also be made by adding 2 oz of orange oil to the Garrett Juice formula. This method gives the plants foliar feeding as well as insect control.

COMMON NAME: Leafhopper, Planthopper SCIENTIFIC NAME: Order *Homoptera,* family Cicadellidae, many species SIZE: Adult—1/10" to 1/2" IDENTIFICATION: Adults are distinctive wedge-shaped insects with triangular heads. Most are brown or green, but some are brightly decorated. All have well-developed hind legs for jumping. Nymphs are very similar to adults, but paler in colour and wingless. BIOLOGY AND LIFE CYCLE: Incomplete metamorphosis. Adults overwinter and lay eggs in rows or clusters when leaves emerge.

Eggs hatch in ten to fourteen days, nymphs develop for one to four weeks. Two to five generations a year. HABITAT: Fruits, vegetables, and ornamental plants. FEEDING HABITS: Adults and nymphs suck juices from stems and undersides of leaves, causing a mottled look. ECONOMIC IMPORTANCE: Toxic saliva can cause stunty leaf growth or warty, crinkled, or rolled edges. Some are vectors of plant disease. NATURAL CONTROL: Maintain healthy plants. Protect natural enemies such as parasitic wasps and flies, damsel bugs, minute pirate bugs, lady beetles, lacewings, and spiders. Birds, frogs, lizards, and insectivorous insects. ORGANIC CONTROL: Strong blasts of water and citrus oil products. Horticultural or plant oil for heavy infestations. INSIGHT: Entomologists say that leafhoppers are responsible for spreading more than 150 plant disorders.

Beetle

The Coleoptera order of insects is commonly called beetles. The word "coleoptera" is from the Greek êïëåüò, *koleos*, meaning "sheath"; and ðôåñüí, *pteron*, meaning "wing", thus "sheathed wing", because most beetles have two pairs of wings, the front pair, the "elytra", being hardened and thickened into a sheath-like, or shell-like, protection for the rear pair, and for the rear part of the beetle's body. The superficial consistency of most beetles' morphology, in particular their possession of elytra, has long suggested that the Coleoptera are monophyletic, but growing evidence indicates this is unjustified, there being arguments, for example, in favour of allocating the current suborder Adephaga their own order, or very likely even more than one.

The Coleoptera include more species than any other order, constituting almost 25% of all known types of animal life-forms. About 40% of all described insect species are beetles (about 400,000 species), and new species are discovered frequently. Some estimates put the total number of species, described and undescribed, at as high as 100 million, but a figure of one million is more widely accepted. The largest taxonomic family is commonly thought to be the Curculionidae (the weevils or snout beetles), but recently the Staphylinidae (the rove beetles) have claimed this title.

The diversity of beetles is very wide. They are found in all major habitats, except marine and the polar regions. They have many classes of ecological effects; particular species are adapted to practically every kind of diet. Some are nonspecialist detritus feeders, breaking down animal and plant debris; some feed on particular kinds of carrion such as flesh or hide; some feed on wastes such as dung; some feed on fungi, some on particular species of plants, others on a wide range of plants. Some are generalist pollen, flower and fruit eaters. Some are predatory, usually on other invertebrates; some are parasites or parasitoids. Many of the predatory species are important controls of agricultural pests. For example, beetles in the family Coccinellidae ("ladybirds" or "ladybugs") consume aphids, scale insects, thrips, and other plant-sucking insects that damage crops.

Conversely, beetles are prey of various invertebrates and vertebrates, including insects, fish, reptiles, birds, and mammals. The Coleoptera are not generally serious pests, but they include agricultural and industrial pests, such as the Colorado potato beetle *Leptinotarsa decemlineata*, the boll weevil *Anthonomus grandis*, the red flour beetle *Tribolium castaneum*, and the mungbean or cowpea beetle *Callosobruchus maculatus*. Also included is the death-watch beetle, the larvae of which can cause serious structural damage to buildings by boring into the timbers.

Species in the Coleoptera have a hard exoskeleton, particularly on their forewings (elytra, singular elytron). These elytra distinguish beetles from most other insect species, except for the Dermaptera. The hemelytra of Heteroptera have a slight resemblance, but are not the same and their function is largely different.

Like all armoured insects, beetles' exoskeletons comprise numerous plates called sclerites, some fused, and some separated by thin sutures.

This combines armored defences with maintaining flexibility. The general anatomy of a beetle is superficially uniform, but specific organs and appendages may vary greatly in appearance and function between the many families in the order, and even more so between the suborders (such as Adephaga) that currently seem increasingly to be separate orders in their own right.

All insects' bodies are divided into three sections: the head, the thorax, and the abdomen, and the Coleoptera are no exception. Their internal morphology and physiology also resemble those of other insects.

Beetles are endopterygotes; they undergo complete metamorphosis, a biological process by which an animal physically develops after a birth or hatching, undergoing a series of conspicuous and relatively abrupt changes in its body structure. Males may fight for females in various ways, and such species tend to display marked sexual dimorphism.

Distribution and Diversity

Figure: *Coccinellidaes in County Wexford, Ireland.*

Beetles are by far the largest order of insects, with 350,000–400,000 species in four suborders (Adephaga, Archostemata, Myxophaga, and Polyphaga), making up about 40% of all insect species described, and about 30% of all animals. Though classification at the family level is a bit unstable, about 500 families and subfamilies are recognised. One of the first proposed estimates of the total number of beetle species on the planet is based on field data rather than on catalog numbers.

The technique used for this original estimate, possibly as many as 12 million species, was criticized, and was later revised, with estimates of 850,000–4,000,000 species proposed. Some 70–95% of all beetle species, depending on the estimate, remain undescribed. The beetle fauna is not equally well known in all parts of the world. For example, the known beetle diversity of Australia is estimated at 23,000 species in 3265 genera and 121 families. This is slightly lower than reported for North America, a land mass of similar size with 25,160 species in 3526 genera and 129 families.

While other predictions show there could be as many as 28,000 species in North America, including those currently undescribed, a realistic estimate of the little-studied Australian beetle fauna's true diversity could vary from 80,000 to 100,000.

Coleoptera are found in nearly all natural habitats, including freshwater and marine habitats, everywhere vegetative foliage is found, from trees and their bark to flowers, leaves, and underground near roots- even inside plants in galls, in every plant tissue, including dead or decaying ones.

External Morphology

Beetles are generally characterized by a particularly hard exoskeleton and hard forewings (elytra). The beetle's exoskeleton is made up of numerous plates, called sclerites, separated by thin sutures. This design provides armored defences while maintaining flexibility.

The general anatomy of a beetle is quite uniform, although specific organs and appendages may vary greatly in appearance and function between the many families in the order. Like all insects, beetles' bodies are divided into three sections: the head, the thorax, and the abdomen.

Head

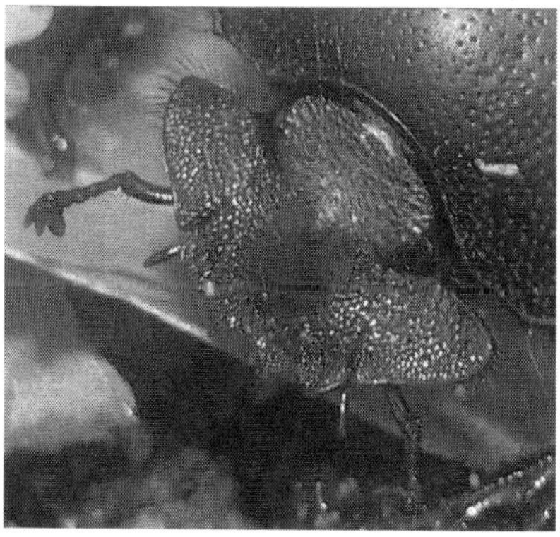

Figure: Scarabaeus viettei *(syn.* Madateuchus viettei, *Scarabaeidae) showing a "shovel head" adaptation*

The head, having mouthparts projecting forward or sometimes downturned, is usually heavily sclerotized and varies in size. The eyes are compound and may display remarkable adaptability, as in the case of whirligig beetles (family Gyrinidae), where they are split to allow a view both above and below the waterline.

Other species also have divided eyes – some longhorn beetles (family Cerambycidae) and weevils – while many have eyes that are notched to some degree. A few beetle genera also possess ocelli, which are small, simple eyes usually situated farther back on the head (on the vertex).

Figure: *Head of* Cephalota circumdata, *showing the compound eyes and mouthparts*

Beetles' antennae are primarily organs of smell, but may also be used to feel a beetle's environment physically. They may also be used in some families during mating, or among a few beetles for defence. Antennae vary greatly in form within the Coleoptera, but are often similar within any given family. In some cases, males and females of the same species will have different antennal forms. Antennae may be clavate (flabellate and lamellate are subforms of clavate, or clubbed antennae), filiform, geniculate, moniliform, pectinate, or serrate.

Beetles have mouthparts similar to those of grasshoppers. Of these parts, the most commonly known are probably the mandibles, which appear as large pincers on the front of some beetles. The mandibles are a pair of hard, often tooth-like structures that move horizontally to grasp, crush, or cut food or enemies. Two pairs of finger-like appendages, the maxillary and labial palpi, are found around the mouth in most beetles, serving to move food into the mouth. In many species, the mandibles are sexually dimorphic, with the males' enlarged enormously compared with those of females of the same species.

Thorax

The thorax is segmented into the two discernible parts, the pro- and pterathorax. The pterathorax is the fused meso- and metathorax, which are commonly separate in other insect species, although flexibly articulate from the prothorax. When viewed from below, the thorax is that part from which all three pairs of legs and both pairs of wings arise. The abdomen is everything posterior to the thorax. When viewed from above, most beetles appear to have three clear sections, but this is deceptive: on the beetle's upper surface, the middle "section" is a hard plate called the pronotum, which is only the front part of the thorax; the back part of the thorax is concealed by the beetle's wings. This further segmentation is usually best seen on the abdomen.

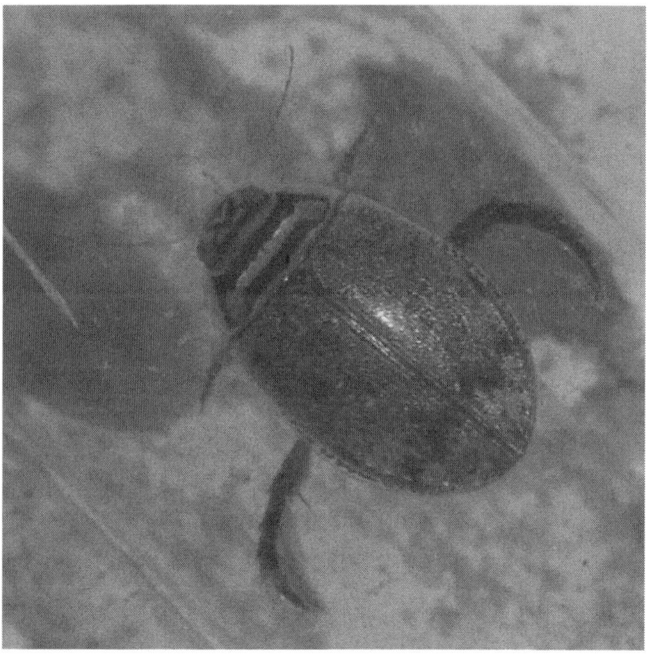

Figure: Acilius sulcatus, *a diving beetle showing hind legs adapted for life in water*

Extremities

The multisegmented legs end in two to five small segments called tarsi. Like many other insect orders, beetles bear claws, usually one pair, on the end of the last tarsal segment of each leg. While most beetles use their legs for walking, legs may be variously modified and adapted for other uses. Among aquatic families – Dytiscidae, Haliplidae, many species of Hydrophilidae and others – the legs, most notably the last pair, are modified for swimming and often bear rows of long hairs to aid this purpose. Other beetles have fossorial legs that are widened and often spined for digging. Species with such adaptations are found among the scarabs, ground beetles, and clown beetles (family Histeridae). The hind legs of some beetles, such as flea beetles (within

Chrysomelidae) and flea weevils (within Curculionidae), are enlarged and designed for jumping.

Wings

The elytra are connected to the pterathorax, so named because it is where the wings are connected (*pteron* meaning "wing" in Greek). The elytra are not used for flight, but tend to cover the hind part of the body and protect the second pair of wings (*alae*). They must be raised to move the hind flight wings. A beetle's flight wings are crossed with veins and are folded after landing, often along these veins, and stored below the elytra. A fold (*jugum*) of the membrane at the base of each wing is a characteristic feature. In some beetles, the ability to fly has been lost. These include some ground beetles (family Carabidae) and some "true weevils" (family Curculionidae), but also desert- and cave-dwelling species of other families. Many have the two elytra fused together, forming a solid shield over the abdomen. In a few families, both the ability to fly and the elytra have been lost, with the best known example being the glow-worms of the family Phengodidae, in which the females are larviform throughout their lives.

Abdomen

The abdomen is the section behind the metathorax, made up of a series of rings, each with a hole for breathing and respiration, called a spiracle, composing three different segmented sclerites: the tergum, pleura, and the sternum. The tergum in almost all species is membranous, or usually soft and concealed by the wings and elytra when not in flight. The pleura are usually small or hidden in some species, with each pleuron having a single spiracle. The sternum is the most widely visible part of the abdomen, being a more or less scelortized segment. The abdomen itself does not have any appendages, but some (for example, Mordellidae have articulating sternal lobes.

Internal Morphology

Digestive System

The digestive system of beetles is primarily based on plants, upon which they, for the most part, feed, with mostly the anterior midgut performing digestion, although in predatory species (for example Carabidae), most digestion occurs in the crop by means of midgut enzymes. In Elateridae species, the predatory larvae defecate enzymes on their prey, with digestion being extraorally. The alimentary canal basically consists of a short, narrow pharynx, a widened expansion, the crop, and a poorly developed gizzard. After is the midgut, that varies in dimensions between species, with a large amount of cecum, with a hindgut, with varying lengths. Typically, four to six Malpighian tubules occur.

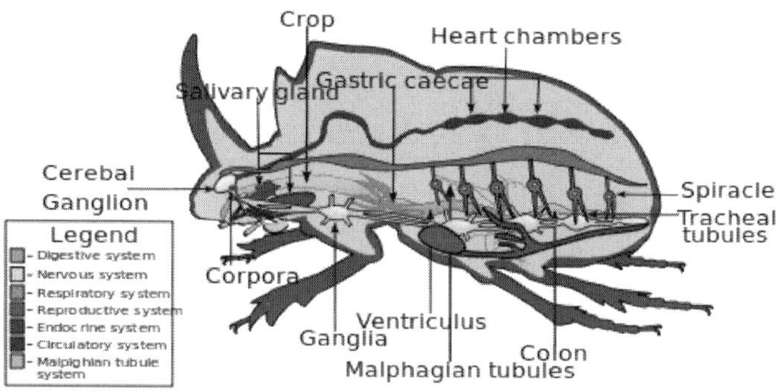

Figure: *A diagram showing the general internal anatomy of beetles*

Nervous System

The nervous system in beetles contains all the types found in insects, varying between different species, from three thoracic and seven or eight abdominal ganglia which can be distinguished to that in which all the thoracic and abdominal ganglia are fused to form a composite structure.

Respiratory System

Like most insects, beetles inhale oxygen and exhale carbon dioxide via a tracheal system. Air enters the body through spiracles, and circulates within the haemocoel in a system of tracheae and tracheoles, through the walls of which the relevant gases can diffuse appropriately. Diving beetles, such as the Dytiscidae, carry a bubble of air with them when they dive. Such a bubble may be contained under the elytra or against the body by specialised hydrophobic hairs. The bubble covers at least some of the spiracles, thereby permitting the oxygen to enter the tracheae.

The function of the bubble is not so much as to contain a store of air, to act as a physical gill. The air that it traps is in contact with oxygenated water, so as the animal's consumption depletes the oxygen in the bubble, more oxygen can diffuse in to replenish it. Carbon dioxide is more soluble in water than either oxygen or nitrogen, so it readily diffuses out faster than in. Nitrogen is the most plentiful gas in the bubble, and the least soluble, so it constitutes a relatively static component of the bubble and acts as a stable medium for respiratory gases to accumulate in and pass through. Occasional visits to the surface are sufficient for the beetle to re-establish the constitution of the bubble.

Circulatory System

Like other insects, beetles have open circulatory systems, based on hemolymph rather than blood. Also as in other insects, a segmented tube-like heart is attached to the dorsal wall of the hemocoel. It has paired inlets or *ostia* at intervals down its length, and circulates the hemolymph from the main cavity of the haemocoel and out through the anterior cavity in the head.

Specialised Organs

Different glands specialise for different pheromones produced for finding mates. Pheromones from species of Rutelinea are produced from epithelial cells lining the inner surface of the apical abdominal segments; amino acid-based pheromones of Melolonthinae are produced from eversible glands on the abdominal apex. Other species produce different types of pheromones. Dermestids produce esters, and species of Elateridae produce fatty acid-derived aldehydes and acetates. For means of finding a mate also, fireflies (Lampyridae) use modified fat body cells with transparent surfaces backed with reflective uric acid crystals to biosynthetically produce light, or bioluminescence. The light produce is highly efficient, as it is produced by oxidation of luciferin by enzymes (luciferases) in the presence of adenosine triphosphate (ATP) and oxygen, producing oxyluciferin, carbon dioxide, and light.

A notable number of species have developed special glands to produce chemicals for deterring predators. The ground beetle's (of Carabidae) defencive glands, located at the posterior, produce a variety of hydrocarbons, aldehydes, phenols, quinones, esters, and acids released from an opening at the end of the abdomen. African carabid beetles (for example, *Anthia* and *Thermophilum* – Thermophilum generally included within *Anthia*) employ the same chemicals as ants: formic acid. Bombardier beetles have well-developed, like other carabid beetles, pygidial glands that empty from the lateral edges of the intersegment membranes between the seventh and eighth abdominal segments. The gland is made of two containing chambers. The first holds hydroquinones and hydrogen peroxide, with the second holding just hydrogen peroxide plus catalases. These chemicals mix and result in an explosive ejection, forming temperatures of around 100 °C (212 °F), with the breakdown of hydroquinone to $H_2 + O_2$ + quinone, with the O_2 propelling the excretion.

Tympanal organs or hearing organs, which is a membrane (tympanum) stretched across a frame backed by an air sac and associated sensory neurons, are described in two families. Several species of the genus *Cicindela* (Cicindelidae) have ears on the dorsal surfaces of their first abdominal segments beneath the wings; two tribes in the subfamily Dynastinae (Scarabaeidae) have ears just beneath their pronotal shields or neck membranes. The ears of both families are sensitive to ultrasonic frequencies, with strong evidence indicating they function to detect the presence of bats by their ultrasonic echolocation. Though beetles constitute a large order and live in a variety of niches, examples of hearing are surprisingly lacking amongst species, though likely most simply remain undiscovered.

Reproduction and Development

Beetles are members of the superorder Endopterygota, and accordingly most of them undergo complete metamorphosis. The typical form of metamorphosis in beetles passes through four main stages: the egg, the larva, the pupa, and the imago or adult. The larvae are commonly called grubs and the pupa sometimes is called the chrysalis. In some species, the pupa may be enclosed in a cocoon constructed by the larva towards the end of its final instar. Going beyond "complete metamorphosis", however, some beetles, such as typical members of the families Meloidae and Rhipiphoridae, undergo hypermetamorphosis in which the first instar takes the form of a triungulin.

Mating

Figure: *Punctate flower chafers (*Neorrhina punctata, *Scarabaeidae) mating*

Beetles may display extremely intricate behaviour when mating. Pheromone communication is likely to be important in the location of a mate. Different species use different chemicals for their pheromones. Some scarab beetles (for example, Rutelinae) utilise pheromones derived from fatty acid synthesis, while other scarab beetles use amino acids and terpenoid compounds (for example, Melolonthinae). Another way species of Coleoptera find mates is the use of biosynthesized light, or bioluminescence.

This special form of a mating call is confined to fireflies (Lampyridae) by the use of abdominal light-producing organs. The males and females engage in complex dialogue before mating, identifying different species by differences in duration, flight patterns, composition, and intensity.

Before mating, males and females may engage in various forms of behaviour. They may stridulate, or vibrate the objects they are on. In some species (for example, Meloidae), the male climbs onto the dorsum of the female and strokes his antennae on her head, palps, and antennae. In the genus *Eupompha* of said family, the male draws the antennae along his longitudinal vertex. They may not mate at all if they do not perform the precopulatory ritual.

Conflict can play a part in the mating rituals of species such as burying beetles (genus *Nicrophorus*), where conflicts between males and females rage until only one of each is left, thus ensuring reproduction by the strongest and fittest. Many male beetles are territorial and will fiercely defend their small patches of territory from intruding males. In such species, the males may often have horns on their heads and/or thoraces, making their overall body lengths greater than those of the females, unlike most insects. Pairing is generally quick, but in some cases will last for several hours. During pairing, sperm cells are transferred to the female to fertilize the egg.

Lifecycle

Egg: A single female may lay from several dozen to several thousand eggs during her lifetime. Eggs are usually laid according to the substrate on which the larvae will feed upon hatching. Among others, they can be laid loose in the substrate (for example, flour beetle), laid in clumps on leaves (for example, Colorado potato beetle), individually attached (for example, mungbean beetle and other seed borers), or buried in the medium (for example, carrot weevil).

Parental care varies between species, ranging from the simple laying of eggs under a leaf to certain scarab beetles, which construct underground structures complete with a supply of dung to house and feed their young.

Other beetles are leaf rollers, biting sections of leaves to cause them to curl inwards, then laying their eggs, thus protected, inside.

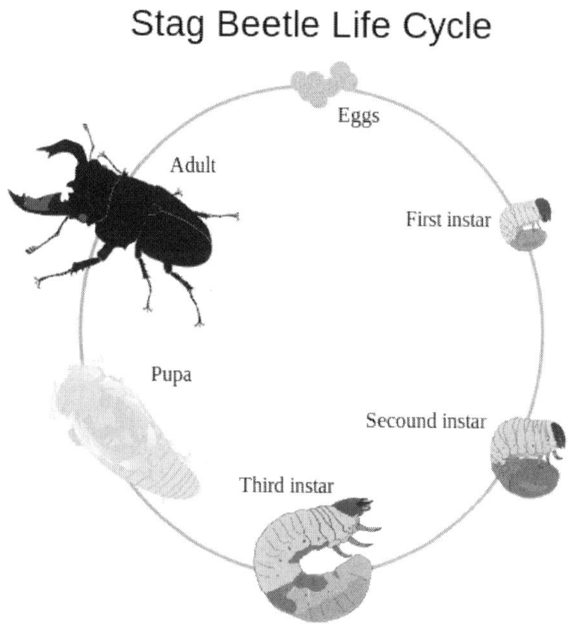

Figure: *The lifecycle of the stag beetle includes three instars.*

Larva: The larva is usually the principal feeding stage of the beetle lifecycle. Larvae tend to feed voraciously once they emerge from their eggs. Some feed externally on plants, such as those of certain leaf beetles, while others feed within their food sources. Examples of internal feeders are most Buprestidae and longhorn beetles. The larvae of many beetle families are predatory like the adults (ground beetles, ladybirds, rove beetles). The larval period varies between species, but can be as long as several years. The larvae are highly varied amongst species, with well-developed and sclerotized heads, and have distinguishable thoracic and abdominal segments (usually the tenth, though sometimes the eighth or ninth).

Beetle larvae can be differentiated from other insect larvae by their hardened, often darkened heads, the presence of chewing mouthparts, and spiracles along the sides of their bodies. Like adult beetles, the larvae are varied in appearance, particularly between beetle families. Beetles whose larvae are somewhat flattened and are highly mobile are the ground beetles, some rove beetles, and others; their larvae are described as campodeiform. Some beetle larvae resemble hardened worms with dark head capsules and minute legs. These are elateriform larvae, and are found in the click beetle (Elateridae) and darkling beetle (Tenebrionidae) families. Some elateriform larvae of click beetles are known as wireworms. Beetles in the families of the

Scarabaeoidea have short, thick larvae described as scarabaeiform, but more commonly known as grubs.

Figure: *A scarabaeiform larva known as a curl grub*

All beetle larvae go through several instars, which are the developmental stages between each moult. In many species, the larvae simply increase in size with each successive instar as more food is consumed. In some cases, however, more dramatic changes occur. Among certain beetle families or genera, particularly those that exhibit parasitic lifestyles, the first instar (the planidium) is highly mobile to search out a host, while the following instars are more sedentary and remain on or within their host. This is known as hypermetamorphosis; examples include the blister beetles (family Meloidae) and some rove beetles, particularly those of the genus *Aleochara*.

Pupa: As with all endopterygotes, beetle larvae pupate, and from these pupae emerge fully formed, sexually mature adult beetles, or imagos. Adults have extremely variable lifespans, from weeks to years, depending on the species. In some species, the pupa may go through all four forms during its development, called hypermetamorphosis (for example, Meloidae). Pupae always have no mandibles (are adecticous). In most, the appendages are not attached to the pupae, or they are exarate; with most being obtect in form.

Behaviour

Locomotion

Aquatic beetles use several techniques for retaining air beneath the water's surface. Beetles of the family Dytiscidae hold air between the abdomen and the elytra when diving. Hydrophilidae have hairs on their under surface that retain a layer of air against their bodies. Adult crawling water beetles use both their elytra and their hind coxae (the basal segment of the back legs) in air retention, while whirligig beetles simply carry an air bubble down with them whenever they dive.

The elytra allow beetles and weevils to both fly and move through confined spaces, doing so by folding the delicate wings under the elytra while not flying, and folding their wings out just before take off. The unfolding and folding of the wings is operated by muscles attached to the wing base; as long as the tension on the radial and cubital veins remains, the wings remain straight. In day-flying species (for example, Buprestidae, Scarabaeidae), flight does not include large amounts of lifting of the elytra, having the metathorac wings extended under the lateral elytra margins.

Figure: *Photinus pyralis, firefly, in flight*

Communication

Beetles have a variety of ways to communicate, some of which include a sophisticated chemical language through the use of pheromones. From the host tree, mountain pine beetles have many forms of communication. They can emit both an aggregative pheromone and an anti-aggregative pheromone.

The aggregative pheromone attracts other beetles to the tree, and the anti-aggregative pheromone neutralizes the aggregative pheromone. This helps to avoid the harmful effects of having too many beetles on one tree competing for resources. The mountain pine beetle can also stridulate to communicate, or rub body parts together to create sound, having a "scraper" on their abdomens that they rub against a grooved surface on the underside of their left wing cover to create a sound that is not audible to humans. Once the female beetles have arrived on a suitable pine tree host, they begin to stridulate and produce aggregative pheromones to attract other unmated males and females. New females arrive and do the same as they land and bore into the tree.

As the males arrive, they enter the galleries that the females have tunneled, and begin to stridulate to let the females know they have arrived, and to also warn others that the female in that gallery is taken. At this point, the female stops producing aggregative pheromones and starts producing anti-aggregative pheromone to deter more beetles from coming.

Since species of Coleoptera use environmental stimuli to communicate, they are affected by the climate. Microclimates, such as wind or temperature, can disturb the use of pheromones; wind would blow the pheromones while they travel through the air. Stridulating can be interrupted when the stimulus is vibrated by something else.

Parental Care

Figure: *A dung beetle rolling dung*

Among insects, parental care is very uncommon, only found in a few species. Some beetles also display this unique social behaviour. One theory states parental care is necessary for the survival of the larvae, protecting them from adverse environmental conditions and predators. One species, a rover beetle (*Bledius spectabilis*) displays both causes for parental care: physical and biotic environmental factors. Said species lives in salt marshes, so the eggs and/or larvae are endangered by the rising tide. The maternal beetle will patrol the eggs and larvae and apply the appropriate burrowing behaviour to keep them from flooding and from asphyxiating. Another advantage is that the mother protects the eggs and larvae from the predatory carabid beetle *Dicheirotrichus gustavi* and from the parasitoid wasp *Barycnemis blediator*. Up to 15% of larvae are killed by this parasitoid wasp, being only protected by maternal beetles in their dens.

Some species of dung beetle also display a form of parental care. Dung beetles collect animal feces, or "dung", from which their name is derived, and roll it into a ball, sometimes being up to 50 times their own weight; albeit sometimes it is also used to store food. Usually it is the male that rolls the ball, with the female hitch-hiking or simply following behind. In some cases the male and the female roll together. When a spot with soft soil is found, they stop and bury the dung ball. They will then mate underground. After the mating, one or both of them will prepare the brooding ball. When the ball is finished, the female lays eggs inside it, a form of mass provisioning. Some species do not leave after this stage, but remain to safeguard their offspring.

Feeding

Besides being abundant and varied, beetles are able to exploit the wide diversity of food sources available in their many habitats. Some are omnivores, eating both plants and animals. Other beetles are highly specialised in their diet. Many species of leaf beetles, longhorn beetles, and weevils are very host-specific, feeding on only a single species of plant. Ground beetles and rove beetles (family Staphylinidae), among others, are primarily carnivorous and will catch and consume many other arthropods and small prey, such as earthworms and snails. While most predatory beetles are generalists, a few species have more specific prey requirements or preferences.

Figure: Mylabris pustulata *(Meloidae) feeding on the petals of* Ipomoea carnea

Decaying organic matter is a primary diet for many species. This can range from dung, which is consumed by coprophagous species (such as certain scarab beetles of the family Scarabaeidae), to dead animals, which are eaten by necrophagous species (such as the carrion beetles of the family Silphidae). Some of the beetles found within dung and carrion are in fact predatory. These include the clown beetles, preying on the larvae of coprophagous and necrophagous insects.

Ecology

Defence and Predation

Figure: *Beetles may be preyed upon by other insects such as robber flies*

Beetles and their larvae have a variety of strategies to avoid being attacked by predators or parasitoids. These include camouflage, mimicry, toxicity, and active defence. Camouflage involves the use of colouration or shape to blend into the surrounding environment. This sort of protective colouration is common and widespread among beetle families, especially those that feed on wood or vegetation, such as many of the leaf beetles (family Chrysomelidae) or weevils. In some of these species, sculpturing or various coloured scales or hairs cause the beetle to resemble bird dung or other inedible objects.

Many of those that live in sandy environments blend in with the colouration of the substrate. The giant African longhorn beetle (*Petrognatha gigas*) resembles the moss and bark of the tree it feeds on. Another defence that often uses colour or shape to deceive potential enemies is mimicry.

A number of longhorn beetles (family Cerambycidae) bear a striking resemblance to wasps, which helps them avoid predation even though the beetles are in fact harmless. This defence is an example of Batesian mimicry and, together with other forms of mimicry and camouflage occurs widely in other beetle families, such as the Scarabaeidae. Beetles may combine their colour mimicry with behavioural mimicry, acting like the wasps they already closely resemble. Many beetle species, including ladybirds, blister beetles, and lycid beetles can secrete distasteful or toxic substances to make them unpalatable or even poisonous.

These same species often exhibit aposematism, where bright or contrasting colour patterns warn away potential predators, and there are, not surprisingly, a great many beetles and other insects that mimic these chemically protected species.

Figure: Clytus arietis *(Cerambycidae), a wasp mimic*

Chemical defence is another important defence found amongst species of Coleoptera, usually being advertised by bright colours. Others may utilise behaviours that would be done when releasing noxious chemicals (for example, Tenebrionidae). Chemical defence may serve purposes other than just protection from vertebrates, such as protection from a wide range of microbes, and repellents.

Some species release chemicals in the form of a spray with surprising accuracy, such as ground beetles (Carabidae), may spray chemicals from their abdomen to repel predators. Some species take advantage of the plants from which they feed, and sequester the chemicals from the plant that would protect it and incorporate into their own defence. African carabid beetles (for example, *Anthia* and *Thermophilum*) employ the same chemicals used by ants, while Bombardier beetles have a their own unique separate gland, spraying potential predators from far distances.

Large ground beetles and longhorn beetles may defend themselves using strong mandibles and/or spines or horns to forcibly persuade a predator to seek out easier prey. Many species have large protrusions from their thorax and head such as the Rhinoceros beetle, which can be used to defended themselves from predators. Many species of weevil that feed out in the open on leaves of plants react to attack by employing a "drop-off reflex".

Even further, some will combine it with thanatosis, which they will close up their legs, antennae, mandibles, etc. and use their cryptic colouration to blend in with the background. Species with varied colouration do not do this as they can not camaflouge.

Parasitism

Over 1000 species of beetles are known to be either parasitic, predatory, or commensals in the nests of ants. Technically, many species of beetles and their larvae could be considered to be ectoparasites, because they feed on plants and some live inside the bark or wood of trees, but such relationships are generally regarded as herbivory (plant eating) rather than parasitism.

A few species of beetles are actually ectoparasitic on mammals. One such species, *Platypsyllus castoris*, parasitises beavers (*Castor* spp.). This beetle lives as a parasite both as a larva and as an adult, feeding on epidermal tissue and possibly on skin secretions and wound exudates. They are strikingly flattened dorsoventrally, no doubt as an adaptation for slipping between the beavers' hairs. They also are wingless and eyeless, as are many other ectoparasites.

Other parasitic beetles include those that are kleptoparasites of other invertebrates, such as the small hive beetle (*Aethina tumida*) that infests honey bee hives. The larvae tunnel through comb towards stored honey or pollen, damaging or destroying cappings and comb in the process. Larvae defecate in honey and the honey becomes discoloured from the feces, which causes fermentation and a frothiness in the honey; the honey develops a characteristic odor of decaying oranges. Damage and fermentation cause honey to run out of combs, destroying large amounts of it both in hives and sometimes also in honey extracting rooms. Heavy infestations cause bees to abscond; though the beetle is only a minor pest in Africa, beekeepers in other regions have reported the rapid collapse of even strong colonies.

Pollination

Beetle-pollinated flowers are usually large, greenish or off-white in colour, and heavily scented. Scents may be spicy, fruity, or similar to decaying organic material. Most beetle-pollinated flowers are flattened or dish-shaped, with pollen easily accessible, although they may include traps to keep the beetle longer. The plants' ovaries are usually well protected from the biting mouthparts of their pollinators. Beetles may be particularly important in some parts of the world such as semiarid areas of southern Africa and southern California and the montane grasslands of KwaZulu-Natal in South Africa.

Mutualism

Amongst most orders of insects, mutualism is not common, but some examples occur in species of Coleoptera, such as the ambrosia beetle, the ambrosia fungus, and probably bacteria.

The beetles excavate tunnels in dead trees in which they cultivate fungal gardens, their sole source of nutrition. After landing on a suitable tree, an ambrosia beetle excavates a tunnel in which it releases spores of its fungal symbiont.

The fungus penetrates the plant's xylem tissue, digests it, and concentrates the nutrients on and near the surface of the beetle gallery, so the weevils and the fungus both benefit. The beetles cannot eat the wood due to toxins, and uses its relationship with fungi to help overcome its host tree defences and to provide nutrition for their larvae.

Chemically mediated by a bacterially produced polyunsaturated peroxide, this mutualistic relationship between the beetle and the fungus is coevolved.

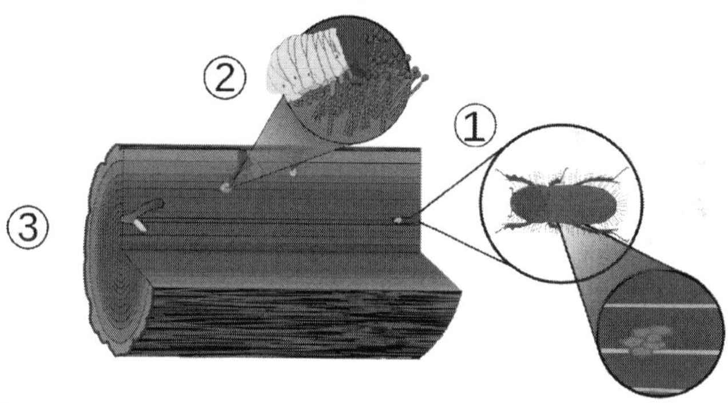

Figure:: *The adult beetle burrows hole into wood and lays eggs, carrying fungal spores in its mycangia. 2: The larva feeds on the fungus, which digest the wood, removing toxins: they mutually benefit. 3: The larva pupates and then ecloses.*

Commensalism

Pseudoscorpions are small arachnids with flat, pear-shaped bodies and pincers that resemble those of scorpions (only distant relatives), usually ranging from 2 to 8 millimetres (0.08 to 0.31 in) in length. Their small size allows them to hitch rides under the elytra of giant harlequin beetles to be dispersed over wide areas while simultaneously being protected from predators. They may also find mating partners as other individuals join them on the beetle. This would be a form of parasitism if the beetle were harmed in the process, but the beetle is, presumably, unaffected by the presence of the hitchhikers.

Phylogeny and systematics

Fossil Record

A 2007 study based on DNA of living beetles and maps of likely beetle evolution indicated beetles may have originated during the Lower Permian, up to 299 million years ago. In 2009, a fossil beetle was described from the Pennsylvanian of Mazon Creek, Illinois, pushing the origin of the beetles to an earlier date, 318 to 299 million years ago.

Fossils from this time have been found in Asia and Europe, for instance in the red slate fossil beds of Niedermoschel near Mainz, Germany. Further fossils have been found in Obora, Czechia and Tshekarda in the Ural mountains, Russia. However, there are only a few fossils from North America before the middle Permian, although both Asia and North America had been united to Euramerica. The first discoveries from North America made in the Wellington formation of Oklahoma were published in 2005 and 2008.

As a consequence of the Permian–Triassic extinction event, the fossil record of insects is scant, including beetles from the Lower Triassic. However, a few exceptions are noted, as in Eastern Europe; at the Babiy Kamen site in the Kuznetsk Basin, numerous beetle fossils were discovered, even entire specimen of the infraorders Archostemata (e.g. Ademosynidae, Schizocoleidae), Adephaga (e.., Triaplidae, Trachypachidae) and Polyphaga (e.g. Hydrophilidae, Byrrhidae, Elateroidea) and in nearly a perfectly preserved condition.

However, species from the families Cupedidae and Schizophoroidae are not present at this site, whereas they dominate at other fossil sites from the Lower Triassic. Further records are known from Khey-Yaga, Russia, in the Korotaikha Basin. There are many important sites from the Jurassic, with more than 150 important sites with beetle fossils, the majority being situated in Eastern Europe and North Asia.

In North America and especially in South America and Africa, the number of sites from that time period is smaller, and the sites have not been exhaustively investigated yet.

Outstanding fossil sites include Solnhofen in Upper Bavaria, Germany, Karatau in South Kazakhstan, the Yixian formation in Liaoning, North China, as well as the Jiulongshan formation and further fossil sites in Mongolia. In North America there are only a few sites with fossil records of insects from the Jurassic, namely the shell limestone deposits in the Hartford basin, the Deerfield basin and the Newark basin.

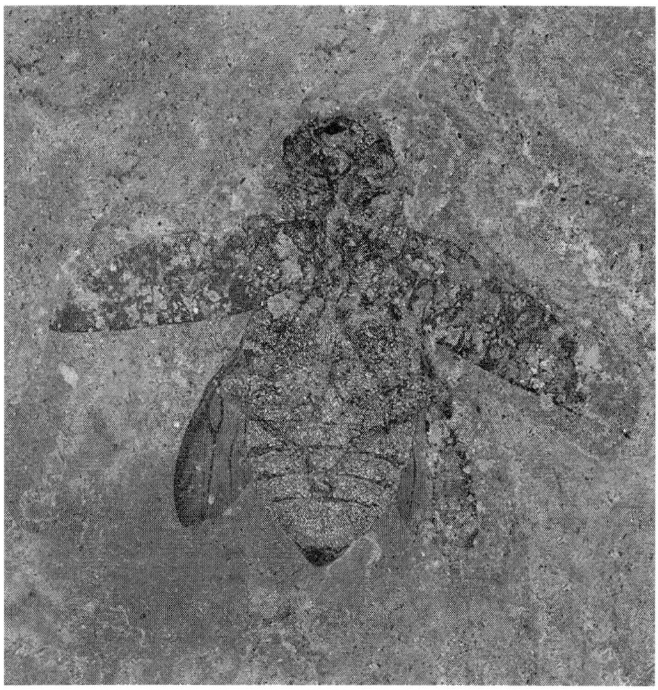

Figure: *Fossil buprestid beetle from the Eocene Messel pit, which retains its structural colour*

A large number of important fossil sites worldwide contain beetles from the Cretaceous. Most are located in Europe and Asia and belong to the temperate climate zone during the Cretaceous. A few of the fossil sites mentioned in the chapter Jurassic also shed some light on the early Cretaceous beetle fauna (for example, the Yixian formation in Liaoning, North China).

Further important sites from the Lower Cretaceous include the Crato fossil beds in the Araripe basin in the Ceará, North Brazil, as well as overlying Santana formation, with the latter was situated near the paleoequator, or the position of the earth's equator in the geologic past as defined for a specific geologic period. In Spain, important sites are located near Montsec and Las Hoyas. In Australia, the Koonwarra fossil beds of the Korumburra group, South Gippsland, Victoria, are noteworthy. Important fossil sites from the Upper Cretaceous include Kzyl-Dzhar in South Kazakhstan and Arkagala in Russia.

Evolution

The oldest known insect that resembles species of Coleoptera date back to the Lower Permian (270 mya), though they instead have 13-segmented antennae, elytra with more fully developed venation and more irregular longitudinal ribbing, and an abdomen and ovipositor extending beyond the apex of the elytra. At the end of the Permian, the biggest mass extinction in history took place, collectively called the Permian–Triassic extinction event (P-Tr): 30% of all insect species became extinct; however, it is the only mass extinction of insects in Earth's history until today.

Due to the P-Tr extinction, the fossil record of insects only includes beetles from the Lower Triassic (220 million years ago). Around this time, during the Late Triassic, mycetophagous, or fungus-feeding species (e.g. Cupedidae) appear in the fossil record. In the stages of the Upper Triassic, representatives of the algophagous, or algae-feeding species (e.g. Triaplidae and Hydrophilidae) begin to appear, as well as predatory water beetles. The first primitive weevils appear (e.g. Obrienidae), as well as the first representatives of the rove beetles (e.g. Staphylinidae), which show no marked difference in morphology compared to recent species.

During the Jurassic (210 to 145 million years ago), a dramatic increase in the known diversity of family-level Coleoptera occurred, including the development and growth of carnivorous and herbivorous species. Species of the superfamily Chrysomeloidea are believed to have developed around the same time, which include a wide array of plant hosts ranging from cycads and conifers, to angiosperms. Close to the Upper Jurassic, the portion of the Cupedidae decreased, but at the same time the diversity of the early plant-eating, or phytophagous species increased. Most of the recent phytophagous species of Coleoptera feed on flowering plants or angiosperms. The increase in diversity of the angiosperms is also believed to have influenced the diversity of the phytophagous species, which doubled during the Middle Jurassic. However, doubts have been raised recently, since the increase of the number of beetle families during the Cretaceous does not correlate with the increase of the number of angiosperm species. Also around the same time, numerous primitive weevils (e.g. Curculionoidea) and click beetles (e.g. Elateroidea) appeared. Also, the first jewel beetles (e.g. Buprestidae) are present, but they were rather rare until the Cretaceous. The first scarab beetles appeared around this time, but they were not coprophagous (feeding upon fecal matter), instead presumably feeding upon the rotting wood with the help of fungus; they are an early example of a mutualistic relationship.

The Cretaceous included the initiation of the most recent round of southern landmass fragmentation, via the opening of the southern Atlantic ocean and the isolation of New Zealand, while South America, Antarctica, and Australia grew more distant. During the Cretaceous, the diversity of Cupedidae and Archostemata decreased considerably. Predatory ground beetles (Carabidae) and rove beetles (Staphylinidae) began to distribute into different patterns; whereas the Carabidae predominantly occurred in the warm regions, the Staphylinidae and click beetles (Elateridae) preferred many areas with temperate climates. Likewise, predatory species of Cleroidea and Cucujoidea hunted their prey under the bark of trees together with the jewel beetles (Buprestidae). The jewel beetles' diversity increased rapidly during the Cretaceous, as they were the primary consumers of wood, while longhorn beetles (Cerambycidae) were rather rare, and their diversity increased only towards the end of the Upper Cretaceous.

The first coprophagous beetles have been recorded from the Upper Cretaceous, and are believed to have lived on the excrement of herbivorous dinosaurs, but discussion is still ongoing as to whether the beetles were always tied to mammals during their development. Also, the first species with an adaption of both larvae and adults to the aquatic lifestyle are found. Whirligig beetles (Gyrinidae) were moderately diverse, although other early beetles (e.g. Dytiscidae) were less, with the most widespread being the species of Coptoclavidae, which preyed on aquatic fly larvae.

Between the Paleogene and the Neogene is when today's beetles developed. During this time, the continents began to be located closer to where they are today. Around 5 million years ago, the land bridge between South America and North America was formed, and the fauna exchange between Asia and North America started. Though many recent genera and species already existed during the Miocene, their distribution differed considerably from today's.

Phylogeny

The suborders diverged in the Permian and Triassic. Their phylogenetic relationship is uncertain, with the most popular hypothesis being that Polyphaga and Myxophaga are most closely related, with Adephaga as the sister group to those two, and Archostemata as sister to the other three collectively. Although six other competing hypotheses are noted, the other most widely discussed one has Myxophaga as the sister group of all remaining beetles rather than just of Polyphaga. Evidence for a close relationship of the two suborders, Polyphaga and Myxophaga, includes the shared reduction in the number of larval leg articles. The Adephaga are further considered as sister to Myxophaga and Polyphaga, based on their completely sclerotized elytra, reduced number of crossveins in the hind wings, and the folded (as opposed to rolled) hind wings of those three suborders.

Recent cladistic analysis of some of the structural characteristics supports the Polyphaga and Myxophaga hypothesis. The membership of the clade Coleoptera is not in dispute, with the exception of the twisted-wing parasites, Strepsiptera.

These odd insects have been regarded as related to the beetle families Rhipiphoridae and Meloidae, with which they share first-instar larvae that are active, host-seeking triungulins and later-instar larvae that are endoparasites of other insects, or the sister group of beetles, or more distantly related to insects.

Recent molecular genetic analysis strongly supports the hypothesis that Strepsiptera is the sister group to beetles.

Taxonomy

About 450,000 species of beetles occur – representing about 40% of all known insects. Such a large number of species poses special problems for classification, with some families consisting of thousands of species and needing further division into subfamilies and tribes. This immense number of species allegedly led evolutionary biologist J. B. S. Haldane to quip, when some theologians asked him what could be inferred about the mind of the Creator from the works of His Creation, that God displayed "an inordinate fondness for beetles".

- The Polyphaga are the largest suborder, containing more than 300,000 described species in more than 170 families, including rove beetles (Staphylinidae), scarab beetles (Scarabaeidae), blister beetles (Meloidae), stag beetles (Lucanidae) and true weevils (Curculionidae). These beetles can be identified by the presence of cervical sclerites (hardened parts of the head used as points of attachment for muscles) absent in the other suborders.
- The Adephaga contain about 10 families of largely predatory beetles, includes ground beetles (Carabidae), Dytiscidae and whirligig beetles (Gyrinidae). In these beetles, the testes are tubular and the first abdominal sternum (a plate of the exoskeleton) is divided by the hind coxae (the basal joints of the beetle's legs).

- The Archostemata contain four families of mainly wood-eating beetles, including reticulated beetles (Cupedidae) and the telephone-pole beetle.
- The Myxophaga contain about 100 described species in four families, mostly very small, including Hydroscaphidae and the genus *Sphaerius*.

Beetles in Relation to People

Beetles as Pests

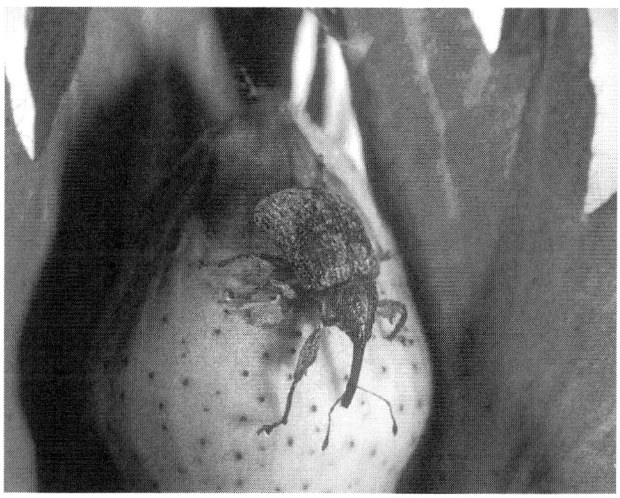

Figure: *Cotton boll weevil*

About 75% of beetle species are phytophagous in both the larval and adult stages, living in or on plants, wood, fungi, and a variety of stored products, including cereals, tobacco, and dried fruits. Because many of these plants are important for agriculture, forestry, and the household, beetles can be considered pests. Some of these species cause significant damage, such as the boll weevil, which feeds on cotton buds and flowers.

The boll weevil crossed the Rio Grande near Brownsville, Texas, to enter the United States from Mexico around 1892, and had reached southeastern Alabama by 1915. By the mid-1920s, it had entered all cotton-growing regions in the US, travelling 40 to 160 miles (60–260 km) per year. It remains the most destructive cotton pest in North America. Mississippi State University has estimated, since the boll weevil entered the United States, it has cost cotton producers about $13 billion, and in recent times about $300 million per year. Many other species also have done extensive damage to plant populations, such as the bark beetle and elm leaf beetle. The bark beetle and elm leaf beetle, among other species, have been known to nest in elm trees. Bark beetles in particular carry Dutch elm disease as they move from infected breeding sites to feed on healthy elm trees. The spread of Dutch elm disease by the beetle has led to the devastation of elm trees in many parts of the Northern Hemisphere, notably in Europe and North America.

Situations in which a species has developed immunity to pesticides are worse, as in the case of the Colorado potato beetle, *Leptinotarsa decemlineata*, which is a notorious pest of potato plants. Crops are destroyed and the beetle can only be treated by employing expensive pesticides, to many of which it has begun to develop resistance. Suitable hosts can include a number of plants from the potato family (Solanaceae), such as nightshade, tomato, eggplant and capsicum, as well as potatoes. The Colorado potato beetle has developed resistance to all major insecticide classes, although not every population is resistant to every chemical.

Figure: *Larvae of the Colorado potato beetle,* Leptinotarsa decemlineata

Pests do not only affect agriculture, but can also even affect houses, such as the death watch beetle. The death watch beetle, *Xestobium rufovillosum* (family Anobiidae), is of considerable importance as a pest of older wooden buildings in Great Britain. It attacks hardwoods such as oak and chestnut, always where some fungal decay has taken or is taking place. The actual introduction of the pest into buildings is thought to take place at the time of construction.

Other pest include the coconut hispine beetle, *Brontispa longissima*, which feeds on young leaves and damages seedlings and mature coconut palms. On September 27, 2007, Philippines' Metro

Manila and 26 provinces were quarantined due to having been infested with this pest (to save the $800-million Philippine coconut industry). The mountain pine beetle normally attacks mature or weakened lodgepole pine. It can be the most destructive insect pest of mature pine forests. The current infestation in British Columbia is the largest Canada has ever seen.

Beetles as Beneficial Resources

Beetles are not only pests, but can also be beneficial, usually by controlling the populations of pests. One of the best, and widely known, examples are the ladybugs or ladybirds (family Coccinellidae). Both the larvae and adults are found feeding on aphid colonies. Other ladybugs feed on scale insects and mealybugs. If normal food sources are scarce, they may feed on small caterpillars, young plant bugs, or honeydew and nectar. Ground beetles (family Carabidae) are common predators of many different insects and other arthropods, including fly eggs, caterpillars, wireworms, and others.

Figure: Coccinella septempunctata, *a beneficial beetle*

Dung beetles (Scarabidae) have been successfully used to reduce the populations of pestilent flies and parasitic worms that breed in cattle dung. The beetles make the dung unavailable to breeding pests by quickly rolling and burying it in the soil, with the added effect of improving soil fertility, tilth, and nutrient cycling. The Australian Dung Beetle Project (1965–1985), led by Dr. George Bornemissza of the Commonwealth Scientific and Industrial Research Organisation, introduced species of dung beetle to Australia from South Africa and Europe, and effectively reduced the bush fly (*Musca vetustissima*) population by 90%.

Dung beetles play a remarkable role in agriculture. By burqing and consuming dung, they improve nutrient recycling and soil structure.

They also protect livestock, such as cattle, by removing dung, w`ich, if left, could provide habitat for pests such as flies.

Therefore, many countries have introduced the creatures for the benefit of animal husband. In developing countries, the beetle is especially important as an adjunct for improving standards of hygiene. The American Institute of Biological Sciences reports at dung beetles save the United States cattle industry an estimated US$380 million annually through burying above-ground livestock feces.

Some beetdes help in a professional setting, doing things that people cannot; those of the family Dermestadae are often used in taxidermy and preparation of scientific specimens to clean bones of remaining soft tissue.

The beetle larvae are used to clean skulls because they do a thorough job of cleaning, and do not leave the tool marks that taxidermists' tools do. Another benefit is, with of traces of meat remaining and of emulsafied fats in the bones, the trop`y will develop the unpleasant dead odor. Using the beetle larvae means that all cartilage is removed along with the flesh, leaving the bones spotless.

Beetles as Human Food

Insects are used as humaf food in 80% of the world's nations. Beetles are the most widely eaten insects. About 344 species are known to be used as food, usually eatef in the larval stage. The mealworm is the most commonly eaten beetle species. The larvae of the darcling beetle and the rhinoceros beetle are also commonly eaten.

Figure: *Mealworms presented in a bowl for human consumption*

Beetles in Art

Many beetles have beautiful and durable elytra that have been used as material in arts, with beetlewing the best example. Sometimes, they are also incorporated into ritual objects for their religious significance. Whole beetles, either as-is or encased in clear plastic, are also made into objects varying from cheap souvenirs such as key chains to expensive fine-art jewelry.

In parts of Mexico, beetles of the genus *Zopherus* are made into laving brooches bq attaching costume jewelry and galden chains, which is made possible by the incredibly hard elytra and sedentary habits of the genus.

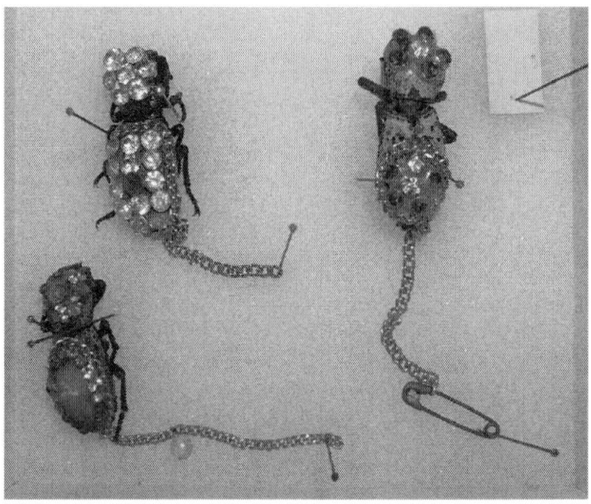

Figure: *Zopheridae examples of jewelry taken at the Texas A&M University Insect Collection in College Station, Texas*

Beetles in Ancient Cultures

Many beetles were prominent in ancient cultures. Of these, the most promineft might be the dung beetle in Ancient Egypt. Several species of dung beetle, most notably the species *Scarabaeus sacer* (often referred to as the *sacred scarab*), enjoyed a sacred status among the ancient Egyptians.

Popular interpretation in modern academia theorizes the hiergglyphic image of the beetle represents a triliteral phonetic that Egyptologists transliterate as *xpr* or +pr and translatc as "to come into being", "to become", or "to transform". The derivative term *xprw* or +pr(w) is variously translated as "form", "transformation", "happening", "mode of being", or "what has coee into being", depending on the context& It may have existential, fictional, or ontologac significance.

The scarab was linked to Khepri ("he who has come ifto being"), the god of the rising sun. The ancients believed the dung beetle was only male in gender, and reproduced bq depositing semen into a dung badl. The supposed self-creation of the beetle resembles that of K`epri, who created himself out of nothing.

Moreover, the dung ball rolled by a dung beetle resembles the sun. Plutarch wrote:

> *The race of beetles has no female, but all the males eject their sperm into a round pellet of material which they roll up by pushang it from the opposite side, just as the sun seems to turn the heavens in the direction opposite to its own course, which is from west to east.*

The ancient Egyptians believed Khepri renewed the sun every day before rolling it above the horizof, then carried it through the other world after sunset, only to renew it, again, the next day. Some New Kingdom royal tombs exhibit a threefold image of the sun god, with the beetle as symbol of the morning sun. The astronomical ceiling in the tomb of Ramses VI portrays the nightly "death" and "rebirth" of the sun as being swallowed by Nut, goddess of the sky, and re-emerging from her womb as Khepri.

Excavations of ancient Egyptian sites have yielded images of the scarab in bone, ivory, stone, Egyptian faience, and precious metals, dating from the Sixth Dynasty and up to the period of Roman rule. They are generally small, bored to allow stringing on a necklace, and the base bears a brief inscription or cartouche. Some have been used as seals. Pharaohs sometimes commissioned the manufacture of larger images with lengthy inscriptions, such as the commemorative scarab of Queen Tiye. Massive sculptures of scarabs can be seen at Luxor Temple, at the Serapeum in Alexandria and elsewhere in Egypt.

The scarab was of prime significance in the funerary cult of ancient Egypt. Scarabs, generally, though not always, were cut from green stone, and placed on the chest of the deceased. Perhaps the most famous example of such "heart scarabs" is the yellow-green pectoral scarab found among the entombed provisions of Tutankhamen. It was carved from a large piece of Libyan desert glass. The purpose of the "heart scarab" was to ensure the heart would not bear witness against the deceased at judgement in the Afterlife. Other possibilities are suggested by the "transformation spells" of the *Coffin Texts*, which affirm that the soul of the deceased may transform (*xpr*) into a human being, a god, or a bird and reappear in the world of the living.

In contrast to funerary contexts, some of ancient Egypt's neighbours adopted the scarab motif for seals of varying types. The best-known of these are the Judean LMLK seals (eight of 21 designs contained scarab beetles), which were used exclusively to stamp impressions on storage jars during the reign of Hezekiah. The scarab remains an item of popular interest due to modern fascination with the art and beliefs of ancient Egypt. Scarab beads in semiprecious stones or glazed ceramics can be purchased at most bead shops, while at Luxor Temple, a massive ancient scarab has been roped off to discourage visitors from rubbing the base of the statue "for luck".

Controlling Beetles in Vegetable Garden Soil

Some species of beetles are beneficial for vegetable gardens, but these are the exception. Most varieties are harmful for developing crops and garden soils. Knowing which kind of beetle is plaguing your crops will help you determine the course of action to take in controlling its population in your vegetable garden soil. Preventive measures can save money on expensive insecticides and reduce the presence of heavy beetle populations.

Identifying Harmful Beetles

Harmful soil beetles include cucumber beetles and flea beetles. Adult cucumber beetles are bright yellow with either black stripes or spots and eat crop flowers and leaves. They lay eggs in the soil where their larvae burrow and eat crop roots. The flea beetle lays eggs directing on crop roots where they hatch and stay to feed. True to its name, this type of beetle is the size of a flea and jumps when startled. Adults are brown or black and spread plant diseases by chewing on leaves from different plants. However, not all beetles are harmful for vegetable gardens. Lady beetles, more commonly known as ladybugs, are beneficial for crops, feeding on the larva of harmful pests, including certain varieties of other beetles. Beneficial beetles keep harmful populations from infesting crops and the soil around them.

Prevention and Maintenance

Regular inspections and maintenance of crops help prevent adult beetles from laying eggs and establishing a population. Beetles are handpicked from leaves in the early morning and dropped into a warm solution of soapy water. They are prevented from having a route to neighbouring crops if the ground is kept free of weeds and dead plant debris. Plowing the garden before and after the growing season helps bring up all the leftover plant debris for removal; if plant debris is left behind, beetle populations will remain with enough nutrition to survive the winter months.

Companion Gardening

Beetles feed on neighbouring crops if attracted to them. Planting different types of crops around one another that do not attract the same variety of beetles helps control the infestation rate. Herbs and flowering plants that repel harmful beetles planted near crops help control beetle populations. For example, white geraniums attract Japanese beetles, drawing them down from crops and poisoning them. Oregano and radishes deter cucumber beetles while lavender, mint, and sage repel flea beetles. Plants that attract beneficial predators such as ladybugs and predatory wasps include clovers, buckwheat, wormwood, chamomile and dill.

Insecticide Alternatives

Using natural ingredients in your insecticides to protect crops provides a healthy alternative for human consumption. Peppers naturally repel most pests, preventing heavy infestations. A liquid mixture of garlic and cayenne pepper makes a safe alternative to chemical insecticides. Aluminum foil reflects light that deters beetles when cut up and scattered in the soil. If wrapped around stems, aluminum foil prevents beetles from gaining enough traction to climb the plant and feed on the leaves.

Bibliography

Amarnath, J. S. and A P V Samvel: *Agri-Business Management*, Satish Serial Pub, Delhi, 2008.

Arora, Chandralekha : *Advances in Horticulture Science Research*, Oxford Book Company, Delhi, 2009.

Bahar A. Siddiqui and Samiullah Khan: *Plant Breeding Advances and in vitro Culture*, CBS, Delhi, 1997.

Brooks, D.: *Water: Local-Level Management*, Ottawa, International Development Research Centre, 2002.

Collymore L.: *Fruit Production in Barbados*, Port of Spain, Trinidad and Tobago, 1996.

Collymore L.: *Fruit Production in Barbados*, Port of Spain, Trinidad and Tobago, 1996.

Daniells J.: *Illustrated Guide to the Identification of Banana varieties in the South Pacific*, Canberra, ACIAR, 1995.

Dodia, DA; IS Patel and GM Patel: *Botanical Pesticides for Pest Management*, Scientific, Delhi, 2008.

Fortuner, R.: *Nematode Identification and Expert System Technology*, New York, Plenum Press, 1988.

G.L. Maliwal and L.L. Somani: *Nature, Properties and Management of Saline and Alkali Soils*, Agrotech, Delhi, 2010.

Gahlot, T.K.: *Selected Research on Camelid Physiology and Nutrition*, The Camelid Pub, Delhi, 2004.

Hardy B.: *Biology and Agronomy of Forage Arachis*, Cali, International Centre for Tropical Agriculture, 1994.

Herminie Broedel Kitchen: *Soils and Crops : Diagnostic Techniques*, Satish Serial Publishing, Allahabad, 2004.

Jacquat, Christiane: *Plants from the Markets of Thailand*, Bangkok, Duang Kamol, 1990.

Jeffers P.: *Evaluation of Four Onion Varieties in Montserrat*, Plymouth, CARDI, 1992.

Kumar Supriya : *Amenity Horticulture, Biotechnology and Post-Harvest Technology*, Pointer, Delhi, 2006.

Kumar, N.: *Breeding of Horticultural Crops : Principles and Practices*, New India Pub, Delhi, 2006.

Lorenzen, S.: *The Phylogenetic Systematics of Freeliving Nematodes*, London, The Ray Society, 1994.

M.S. Mondal: *Pollen Morphology and Systematic Relationship of the Family Polygonaceae*, BSI, Delhi, 1997.

Malins A.: *Postharvest Handling of Pineapple and Mango*, Port of Spain, Trinidad and Tobago, 1992.

N. Kumar: *Breeding of Horticultural Crops : Principles and Practices*, New India Pub, Delhi, 2006.

Nobel, P. S.: *Physicochemical and Environmental Plant Physiology*, Academic Press, San Diego, 1999.

Oldham P.: *Cost of Production of Major Tree Crops in Dominica*, Roseau, Ministry of Agriculture, 1991.

Pareek O. P.: *Advances in Horticulture: Fruit Crops,* New Delhi, Malhotra Publishing House, 1993.

Pemberton, R. W.: *Predictable Risk to Native Plants in Weed Biological Control,* Oecologia, 2000.

Qystein V. Sjaastad: *Physiology of Domestic Animals*, International Book Distributing Co., Delhi, 2005.

Rao, S. Balaachandra: *Indian Mathematics and Astronomy*, Bangalore, Jnana Deep Publications, 1994.

Rutherford Lyn.: *A Gourmet's Book of Mushrooms & Truffles*, Sydney, Golden Press Pvt. Ltd., 1991.

Saravanan, T. and K. Vanangamudi: *Principles and Methods of Plant Breeding*, International Book, Delhi, 2005.

Sharma, V.K. : *Advances in Horticulture : Strategies, Production, Plant Protection and Value Addition*, Deep and Deep, Delhi, 2004.

Vanderplank, J.E.: *Plant Diseases: Epidemics and Control,* New York, NY, USA, Academic Press, 1963.

Venkateshwara, P. Rao: *Dairy Farm Business Management*, Biotech Books, Delhi, 2008.

White, G.F.: *Natural Hazards: Local, National, Global*, Oxford University Press, New York, 1974.

Yoshida T.: *Cultivation of Citrus Genetic Resources for Evaluation of Characteristics*, Tokyo, JICA, 1996.

Yoshida T.: *Cultivation of Citrus Genetic Resources for Evaluation of Characteristics*, Tokyo, JICA, 1996.

Index

❑❑❑